The Mammalian Alimentary System
A Functional Approach

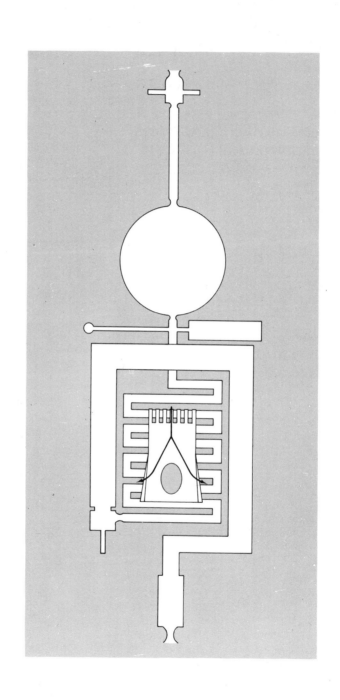

Special Topics in Biology Series

The Mammalian Alimentary System A Functional Approach

David S. Madge
B.Sc., Ph.D., M.I.Biol.
Reader in Zoology, Wye College (University of London)

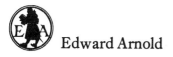 Edward Arnold

First published 1975
by Edward Arnold (Publishers) Limited,
25 Hill Street, London WLX 8LL

Boards Edition ISBN 0 7131 2518 7
Paper Edition ISBN 0 7131 2519 5

Text set in 11 pt. Photon Imprint, printed by photolithography,
and bound in Great Britain at The Pitman Press, Bath

Preface

I'll tell you a story. But if I tell it badly I hope you'll excuse me. I'll do my best; and here's my tale. (Geoffrey Chaucer)

Several authoritative books have recently been published on the structure, functions and malfunctions of the mammalian alimentary system. This work does not attempt to emulate these excellent monographs. Rather, it summarizes the well-known process of extracellular digestion and outlines the remarkable revival of elegant studies on intracellular digestion and absorption in the small intestine.

The expert may be astonished by some of the generalizations, over-simplifications and omissions. But this small book was not written for experts; suffice to state that some of the descriptions given are controversial and await clarification. Yet, "without generalization there is no meaning , and without concreteness there is no significance" (A. N. Whitehead). I apologize for frequently using 'probably' and 'possibly' in the text but for sake of accuracy they cannot be avoided—a tribute, perhaps, to the present state of flux in this story.

This work is aimed mainly at undergraduates and offers an introduction to postgraduates.

Wye and Canterbury, 1974   D.M.

Acknowledgements

To my colleagues who stopped being busy to hold useful discussions with me on diverse topics on the mammalian gut. Also, to the authors and publishers who gave me permission to reproduce the illustrations and poem.

Veronica, Sara, Clare and Philip's

Contents

I Secretion, Digestion and Motility

It ain't so much the things we don't know that gets us in trouble. It's the things we know that ain't so. (Artemus Ward)

1   Introduction

Functions

To remain alive an animal needs food. Mammals obtain food as complex raw materials, salts and vitamins from the external environment and the alimentary canal transfers both food and water to the internal environment where they are distributed to the cells of the body by the circulatory system. W. Prout in 1827 originally divided foodstuffs into "the saccharine, the oily and the albuminous", meaning carbohydrates, fats and proteins respectively. Food is often in the form of large molecules and before these can be utilized they are broken down *(digested)* into small basic units before entering the blood or lymph *(absorbed)*.

Digestion is a process involving many *digestive enzymes* which are secreted into the lumen of the alimentary canal. These enzymes are found in the secretions of the salivary glands, the stomach and a portion of the pancreas. The final stages of digestion of small molecules usually take place within the cells of the small intestine, during which they are absorbed. Some of the enzymes are aided by *acid* from the stomach and *bile* secreted by the liver. Unwanted and unabsorbed materials pass through the lumen of the large intestine and rectum and are ejected as faeces. Large amounts of *fluid* are also secreted by the alimentary canal and its associated organs, most of which are absorbed.

The *muscles* in the alimentary canal wall, often activated by *nerves*, regulate the passage of food and allow time for digestion and absorption. Parts of the inner surface of the alimentary tract secrete minute amounts of *hormones* which are transported in the circulation to profoundly influence the functions in certain regions of the alimentary tract.

Structure

Figure 1.1 illustrates the generalized structure of the mammalian alimentary canal. The alimentary canal is a long, hollow, muscular tube extending from mouth to anus. The wall consists of four concentric layers. Starting from the inside to the outside these are:

1   mucosa, consisting of epithelium, lamina propria and muscularis mucosa,
2   submucosa,
3   muscularis externa, consisting usually of two layers of smooth muscle, and
4   serosa, or adventitious coat.

The lining of the *mucosa* consists of a surface epithelium which varies in different

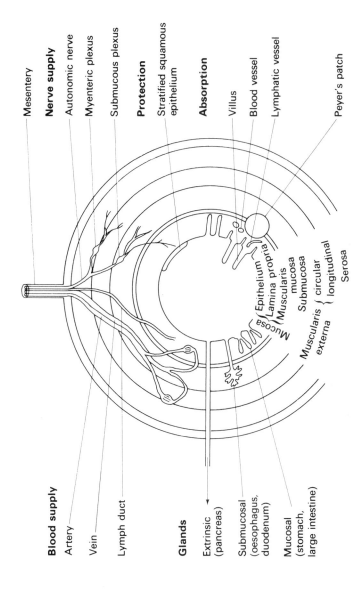

Blood supply

Artery

Vein

Lymph duct

Glands

Extrinsic
(pancreas)

Submucosal
(oesophagus,
duodenum)

Mucosal
(stomach,
large intestine)

Mesentery

Nerve supply

Autonomic nerve

Myenteric plexus

Submucous plexus

Protection

Stratified squamous
epithelium

Absorption

Villus

Blood vessel

Lymphatic vessel

Peyer's patch

Epithelium
Lamina propria
Muscularis
mucosa
Submucosa
Muscularis { circular
externa { longitudinal
Serosa
Mucosa

Fig. 1.1 Generalized structure of the mammalian alimentary canal (after
Passmore and Robson, 1968).

regions of the alimentary canal. It is often glandular and has an absorptive or a secretory surface epithelium. The lamina propria is a loose, underlying connective tissue layer which contains blood capillaries, lymphatic vessels and white blood cells (lymphocytes). The muscularis mucosa is a thin band of smooth muscle fibres, mainly longitudinal muscle.

The *submucosa* is a loose layer of connective tissue. It contains blood capillaries and lymphatic vessels, and a nerve plexus with nerve cells and fibres that innervate both the surface epithelium and the secretory cells.

The *muscularis externa* is composed of an inner circular and outer longitudinal band of smooth muscle. The stomach has an additional overlying layer of smooth, oblique muscle. The anterior region of the oesophagus and terminal region of the anus contain striated muscle fibres. The muscularis externa is innervated by an intricate system of nerve cells and fibres in between the muscle layers.

The *serosa* consists of an outer layer of flattened epithelial (endothelial) cells and a thin layer of loose, underlying connective tissue. The epithelium continues with the *mesentery,* which attaches the intestine to the body wall. It supplies the intestine with nerves, blood vessels and lymphatics.

The intrinsic *blood supply* is particularly evident in regions with absorptive or secretory activities. The arterioles supply capillaries to both the muscular coat and the submucosa. Arterioles also send ascending branches to capillary beds of the muscularis mucosa, glands and lining epithelium. Venules drain these beds into a venous submucous plexus which sends descending branches to the muscular coat.

The *nerve supply* is both intrinsic and extrinsic. Each forms part of the autonomic nervous system, consisting of parasympathetic, sympathetic and visceral afferent fibres. Parasympathetic fibres run into the serosa and enter the muscularis externa where they form the *myenteric plexus of Auerbach.* Other parasypathetic fibres enter the submucosa to form the *submucous plexus of Meissner.* Both plexuses contain parasympathetic ganglionic cells. Symphathetic fibres enter the serosa and innervate blood vessels in between the muscle layers, the muscularis mucosa and glands in the mucosa. In addition, visceral afferent (motor) fibres begin in the mucosa and sometimes in muscle where they act as stretch receptors.

In order to understand how the alimentary system works, its vascular supply, nerves and muscles will be considered in greater detail.

The vascular system

Blood is a liquid tissue, consisting of blood cells (45 per cent) and plasma (55 per cent). It is pumped round the body by the heart through a closed system of blood vessels, forming the systemic (body) and pulmonary (lung) circulation (Fig. 1.2).

Man has five litres of blood in his body which circulates round the body once each minute. Not all the blood is circulating at this rate. Half a litre is in the lungs,

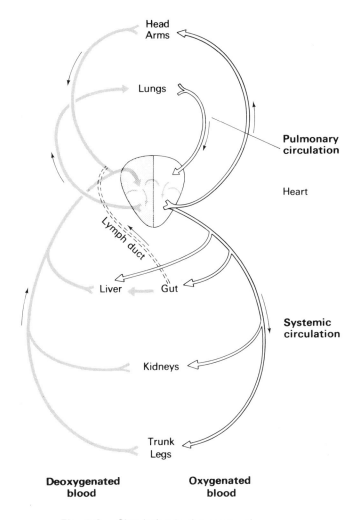

Fig. 1.2   Circulation in the mammal.

one litre in the heart, one litre in the arteries and the remaining two and a half litres in the veins. Blood circulates around the alimentary tract at the rate of one and a half litres per minute. Following a meal, the volume of blood in the alimentary tract rises markedly, owing to hormone action and local reflexes. Secretory and absorptive activities and liver metabolism are then all increased.

Figure 1.3 outlines the blood supply of the alimentary tract. Oxygenated blood

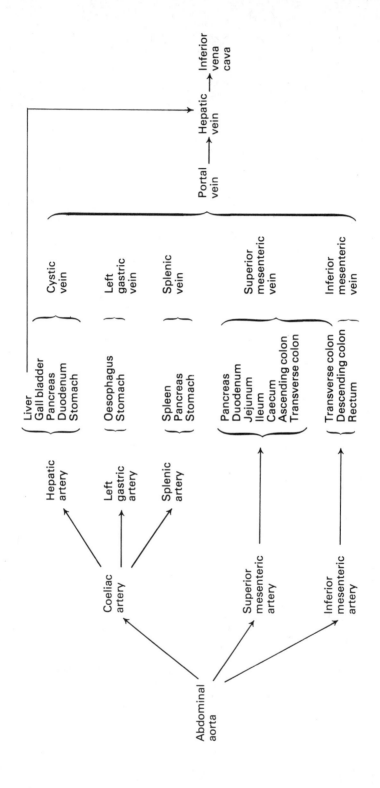

Fig. 1.3   Circulation of the alimentary tract (modified from Texter *et al.*, 1968).

leaves the heart by the dorsal aorta and reaches the abdominal aorta to be distributed to different regions of the alimentary canal and its associated glands by three main arteries. The arteries divide into progressively smaller branches, arterioles, which ultimately ramify into a large number of very small, thin vessels, the capillaries. The capillaries unite to form many small venules, which in turn unite to fewer but larger vessels. the veins. Except for the liver, deoxygenated blood is removed from the alimentary system through several veins into the portal vein which drains into the liver. The liver is drained by the hepatic vein. Venous blood returns to the heart by the inferior vena cava.

Lymph is a colourless fluid containing proteins and varying kinds of cells, mainly lymphocytes. Man has about 150 cm^3 of lymph circulating once each hour by contraction of the surrounding body muscles. Lymphatic capillaries ramify throughout the tissue space and join to form lymphatic vessels. In the alimentary canal wall these vessels unite to form a main lymphatic duct which eventually reaches the heart by the venous blood system. Lymphatic nodes, found at intervals along the lymphatic vessels, filter out and destroy bacteria in the lymph.

The nervous system

The main function of the alimentary canal is to provide the body with food and water. Before it can do this the food must be moved along the alimentary canal and mixed with its secretions. These movements are performed by the muscles in the alimentary tract wall which are often co-ordinated by the nervous system.

MAIN DIVISIONS The *nervous system* consists of a central part, the *brain* and *spinal cord,* linked by a peripheral part, the *nerve fibres.* Nerve fibres form part of nerve cells or *neurones.* Figure 1.4 illustrates the structure of the main types of neurones. The neurones in the brain consist mainly of interneurones. Fibres are either (a) *sensory* or *afferent,* relaying stimuli from receptors in organs and tissues to the central nervous system, or (b) *motor* or *efferent,* relaying information from the central nervous system to effectors in organs and tissues. Afferent fibres enter the spinal cord by the *posterior root* and efferent fibres leave it by the *anterior root.* Soon after leaving the posterior and anterior roots the fibres run together to form *spinal nerves.* There are 31 pairs of spinal nerves and 12 pairs of cranial nerves.

The peripheral system is divided into the somatic nervous system and the autonomic nervous system. The *somatic nervous system* is concerned with the responses of the animal to external stimuli and so with 'voluntary' actions, like movements of striated muscles. Typically, the somatic nervous system involves single afferent and efferent fibres, linked by interneurones.

The *autonomic nervous system* is concerned with the responses of the animal to internal stimuli that are not under voluntary control, like movements of the alimentary canal and its secretions. The autonomic nervous system is divided into a sympathetic and a parasympathetic system (p. 9), each consisting of two

Unipolar neurone (motor or efferent)

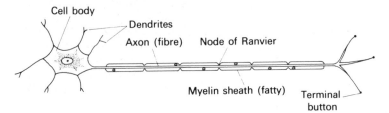

Bipolar neurone (sensory or afferent)

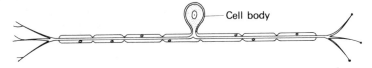

Interneurone

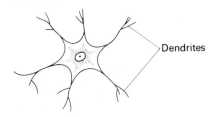

Fig. 1.4   Main types of nerve cells (N.B. the postganglionic fibres of the autonomic nervous system are unmyelinated).

efferent fibres that synapse in a *ganglion*, lying either outside the alimentary canal (sympathetic system) or within its wall (parasympathetic system). The smooth muscles of the gut usually have a mixed spinal nerve supply, containing afferent and efferent, somatic and autonomic fibres.

The autonomic nervous system generally controls movements of the alimentary tract and its secretions. The sympathetic system is inhibitory and the parasympathetic excitatory. Usually the sympathetic system acts as a unit to decrease the activity of the whole alimentary tract, while the parasympathetic system increases the activity in specific regions of the alimentary tract. Sympathetic stimuli also contract the sphincters of the alimentary canal; parasympathetic stimuli relax them. Movements of the alimentary canal and its secretions are often autonomic, determined by intrinsic neuromuscular mechanisms which relay local impules. They are, however, sometimes regulated through extrinsic nerves.

Nerve activity leads to the release of a *chemical transmitter* at the nerve endings (p. 18). When the sympathetic nerves of the alimentary tract are stimulated, *noradrenaline (norepinephrine)* is released at the nerve-muscle junction. When the parasympathetic nerves are stimulated, *acetylcholine* is released. These nerves are called *adrenergic* and *cholinergic* respectively.

Recent work suggests that chemical transmitters other than noradrenaline or acetylcholine may also be released from nerve endings. Non-adrenergic inhibitory nerves have been located in the myenteric plexus that are exclusively intrinsic. Distension of the wall in parts of the alimentary canal releases mainly adenosine triphosphate at the nerve endings, which reflexly inhibits motility. Non-cholinergic excitatory nerves have also been isolated. When either afferent nerve endings or interneurones within the myenteric plexus are stimulated certain chemicals, for example histamine, prostaglandins and serotonin are released which reflexly increase motility.

THE REFLEX ACTION The functional unit of the nervous system is the *reflex action* (Fig. 1.5). Reflex actions occur at all levels of the brain and spinal cord. Very few actions involve only two neurones—the afferent and efferent. Most reflex actions involve at least three neurones. Afferent and efferent neurones are linked by interneurones in the central nervous system, forming very intricate arrays of interacting functional links. Such reflex actions are called *somatic* or *spinal reflexes*.

Reflex actions involving the autonomic nervous system are called *visceral* or *autonomic reflexes*. Both have afferent and efferent fibres. The afferent pathway for both visceral and somatic reflexes is similar, sending stimuli from receptors in the alimentary tract to the spinal cord by the posterior root. The visceral reflex involves two efferent (pre- and postganglionic) neurones synapsing in a ganglion, unlike the somatic reflex action which only involves one (efferent) neurone. The efferent pathway leaves the spinal cord by the anterior root.

A visceral reflex action works like this. The visceral (autonomic) centre in the brain and spinal cord receives sensory information from the alimentary tract by the visceral afferent pathway. In the central nervous system interneurones communicate with the higher autonomic centre in the brain (cerebral cortex) which send impulses to modify activities in the lower autonomic centre (thalamus, hypothalamus and medulla) and spinal cord. Impulses are then conducted to the alimentary system by efferent autonomic pathways.

THE SYMPATHETIC SYSTEM Figure 1.6*a* illustrates a sympathetic nerve fibre (cell). The sympathetic outflow of the alimentary tract is limited to some segments of the thoracic (T1 and T5 to T12) and lumbar (L1 to L4) spinal cord (Fig. 1.7). Short, myelinated (surrounded by myelin, a fatty sheath) preganglionic

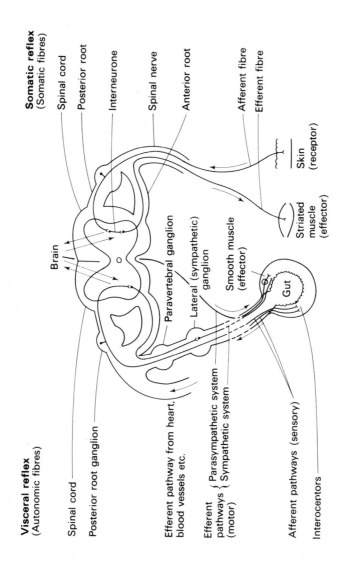

Visceral reflex
(Autonomic fibres)

Spinal cord

Posterior root ganglion

Efferent pathway from heart,
blood vessels etc.

Efferent { Parasympathetic system
pathways { Sympathetic system
(motor)

Afferent pathways (sensory)

Interoceptors

Somatic reflex
(Somatic fibres)

Spinal cord

Posterior root

Interneurone

Spinal nerve

Anterior root

Afferent fibre

Efferent fibre

Skin
(receptor)

Striated
muscle
(effector)

Brain

Paravertebral ganglion

Lateral (sympathetic)
ganglion

Smooth muscle
(effector)

Gut

Fig. 1.5 Visceral and somatic reflex arcs. The preganglionic neurone of the autonomic nervous system is homologous to the interneurone of the somatic nervous system.

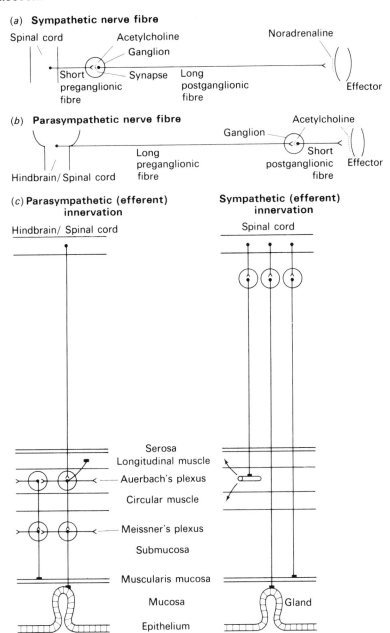

(a) **Sympathetic nerve fibre**

Spinal cord · Acetylcholine · Noradrenaline · Ganglion · Short preganglionic fibre · Synapse · Long postganglionic fibre · Effector

(b) **Parasympathetic nerve fibre**

Ganglion · Acetylcholine · Long preganglionic fibre · Short postganglionic fibre · Effector · Hindbrain/Spinal cord

(c) **Parasympathetic (efferent) innervation** · **Sympathetic (efferent) innervation**

Hindbrain/ Spinal cord · Spinal cord

Serosa
Longitudinal muscle
Auerbach's plexus
Circular muscle
Meissner's plexus
Submucosa
Muscularis mucosa
Mucosa · Gland
Epithelium

Fig. 1.6 Fibres of the autonomic nervous system, and innervation of the alimentary canal. Details in text (modified from Davenport, 1966).

fibres pass from the spinal cord by the anterior nerve root. The fibres pass through, but do not usually communicate with, the sympathetic chain (paravertebral ganglia) running down on either side of the spinal cord, (Figs. 1.5 and 1.7) and innervate the salivary glands, alimentary canal and associated organs. They synapse in four pairs of lateral ganglia situated near the abdominal aorta. The fibres from the superior cervical ganglion (SCG) innervate the three pairs of salivary glands and the oesophagus; those from the coeliac ganglion (CG) innervate the oesophagus, stomach, pancreas and gall bladder; those from the superior mesenteric ganglion (SMG) innervate the stomach, small intestine and large intestine, and those from the inferior mesenteric ganglion (IMG) innervate the large intestine and rectum. The postganglionic (splanchnic) nerves are long and unmyelinated and end in secretory cells in the mucosa of the gut, smooth muscle cells of the muscularis mucosa and in blood vessels among the circular and longitudinal muscle layers, the chemical transmitter diffusing from the blood vessels to the muscle cells (Fig. 1.6c).

THE PARASYMPATHETIC SYSTEM Figure 1.6b illustrates a parasympathetic nerve fibre (cell). Preganglionic fibres are long and myelinated and synapse with short unmyelinated ganglionic fibres in a ganglion within the alimentary tract wall. Parasympathetic fibres associated with the alimentary tract originate as cranial nerves VII, IX and X from the medulla in the hind brain, and from the sacral (S2 to S4) spinal nerves (Fig. 1.7).

Cranial nerves VII and IX innervate the salivary glands. A long preganglionic fibre in the facial nerve (VII) crosses the eardrum as the chorda tympani nerve. A branch meets the (V) lingual nerve going to the tongue and another branch synapses in the submaxillary ganglion (SG), then sends two short postganglionic fibres to the submaxillary (SM) and sublingual (SL) glands. A long preganglionic fibre from the glossopharyngeal nerve (IX) synapses in the otic ganglion (OG), then relays a short postganglionic fibre to the parotid gland (PA).

The vagal (X) nerve is a long preganglionic nerve supplying many organs in the thorax and abdomen, including the alimentary tract as far as the transverse large intestine. The vagus follows blood vessels and synapses in small ganglia in between the circular and longitudinal muscle layers in Auerbach's (myenteric) plexus, then short postganglionic fibres communicate in ganglia with other nerve cells in the submucosa in Meissner's (neurogenic) plexus. Others end in secretory and endocrine cells in the mucosa, the muscularis mucosa and in longitudinal, but not circular, muscle cells of the muscular coat (Fig. 1.6c).

Long preganglionic fibres also arise from the sacral segments (S2 to S4) of the spinal cord and form the pelvic (sacral) nerves. They supply the descending limb of the large intestine, the rectum and anus and, like the vagus, synapse within the wall. Impulses from these fibres are concerned with defaecation.

VISCERAL AFFERENT INNERVATION The autonomic nervous system contains afferent fibres that form part of the visceral (autonomic) reflex arc. Afferent impulses conduct impulses from sensory receptors in the alimentary canal to the central nervous system (Fig. 1.5). Afferent nerves begin in receptors (stretch, pressure, or chemoreceptors) in the mucosal epithelium and in the nerve plexuses within the muscular coat. They travel with the extrinsic innervation and pass through the sympathetic ganglion chain without synapsing, entering the spinal cord by the posterior root where their cell bodies are found, and carry sensations to the central nervous system. The activities are analyzed and integrated by the higher and lower autonomic centres in the brain (p. 9), and impulses are then relayed back to the alimentary tract by efferent fibres of the autonomic nervous system. Visceral afferent fibres mediate visceral (autonomic) reflexes and visceral sensations as, for example, defaecation, pain, hunger and thirst.

Smooth muscle

STRUCTURE The wall of the alimentary canal consists of smooth (involuntary, unstriated) muscle from the oesophagus to the rectum. The upper portion of the oesophagus, however, may consist of striated (voluntary, skeletal) muscle, as does the external sphincter in the anal region. Smooth muscle is composed of long, tapering cells, each 100–150 μm long and up to 10 μm thick (1 μm = 10^{-3} mm), each with a central nucleus, dense bodies, mitochondria, vesicles and a thin surrounding membrane, 0.15 μm thick, containing rows of small vesicles. The cells are packed with contractile myofibrils, parallel to the long axis of the cell. The myofibrils are composed of thin filaments, each 50–80 A thick (1 A = 10^{-4} μm = 10^{-1} nm) consisting of a contracile protein, actin and thick, dark, ribbon-like filaments, each 150–200 Å thick, possibly consisting of another contractile protein, myosin. Adjacent cells are linked together by protuberances (nexuses) of the membrane. Each cell is surrounded by a sheath of fibres, and the cells are arranged in cylindrical bundles separated by connective tissue. The bundles are 50–100 μm in diameter and contain up to 75 muscle cells.

By contrast, striated (voluntary, skeletal) muscle consists of long slender cells up to 50 mm long and 100 μm in diameter, known as muscle fibres, grouped together in bundles. The cells contain myofibrils 2 μm in diameter. Each myofibril is divisible into a series of overlapping filaments, each 150 Å thick, made up of actin and myosin. Much more is known about the structure of striated muscle and how it works, but details are not given here since the musculature of the alimentary canal is mainly smooth.

ELECTROPHYSIOLOGY (MUSCLES AND NERVES) The electrical phenomena for smooth and striated muscle and nerve fibres all broadly follow similar patterns, but much less is known about these events in smooth muscle. Activation of muscle cells depends upon previous electrical events occurring at the cell membrane and in the surrounding fluid (Fig. 1.8). An electrical potential difference exists between both sides of the cell membrane, the interior being negatively

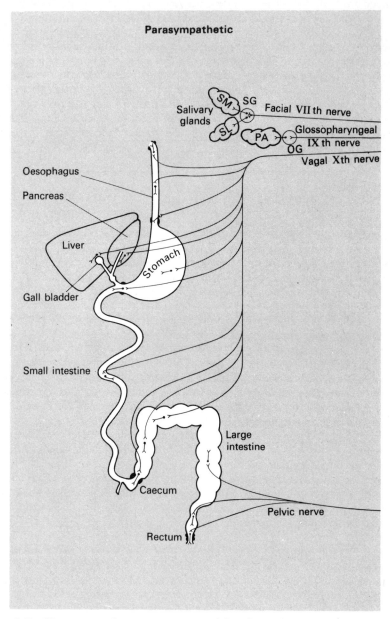

Fig. 1.7   The autonomic nervous system of the alimentary tract. Details in text.

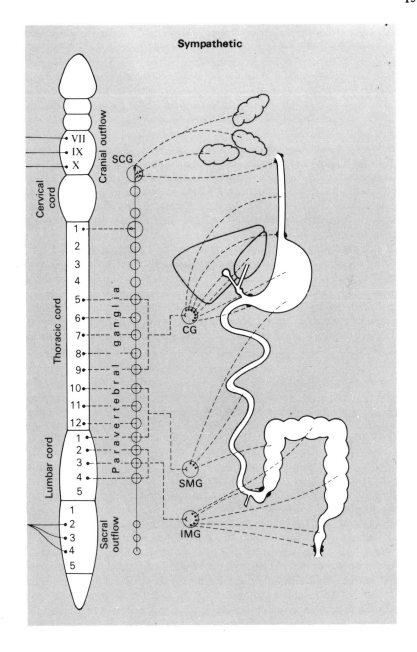

charged and the exterior positively charged. When a smooth muscle cell is inactive, this potential difference, or *resting (membrane) potential,* averages about -60 mV (range -50 to -80 mV). When stimulated, the inside of the cell becomes positive with respect to the exterior, that is, the cell becomes *depolarized.* Beyond a depolarization threshold of about -40 mV the resting potential momentarily reverses to about $+10$ mV, leading to the *action (spike) potential.* As the cell repolarizes, the spike subsides *(negative after potential)* to be followed by a slower phase of afterpolarization *(positive after potential)* before returning to the resting potential.

These electrical changes depend upon the presence of charged ions both in and out of the cell (Fig. 1.8). Muscle cells contain a relatively high concentration of potassium cations and organic (protein) anions and a relatively low concentration of sodium cations; the extracellular fluid is rich in sodium cations and chloride anions. When the cell is stimulated, sodium ions enter the cell passively through the cell membrane, to be quickly pumped out by a sodium pump mechanism. Also, potassium ions flow out passively, followed by a delayed re-entry, and the membrane potential eventually reaches its resting level.

During re-entry, potassium ions are coupled to a sodium pump which is driven by cellular energy (Fig. 1.9). The passive entry of sodium ions causes the action potential. The rapidly falling phase (negative after potential) follows when sodium permeability is falling. The delayed entry of potassium ions causes the positive after potential. The action potential lasts about one ms and after polarization lasts 10 ms or longer. Repeated action potentials lead to muscle contraction.

Although sodium undoubtedly plays an important role in the electrical activity of muscle (and nerve) cells, recent evidence indicates that calcium is at least as important in smooth muscle action. Calcium ions are probably involved in the regulation of membrane permeability, the transfer of current across the membrane during the action potential, and in triggering the contraction process itself.

Visceral smooth muscle is spontaneously active and can contract rythmically and constantly in the absence of nervous stimulation. Striated muscle cannot do this. *Slow waves* originate in the longitudinal muscle layer and become conducted to the circular muscle layer along interconnecting membrane protuberances. Any one longitudinal muscle cell can initiate contraction by acting as a 'pacemaker'. *Pacemaker (pacesetter) potentials* of 1.5 mV or less become rapidly propagated from one cell to another, probably through touching cell membranes, so that the entire smooth muscle layer behaves as a syncytial unit. Pacemaker potentials are periodically superimposed by rapid bursts of action potentials, each burst accompanying a slow wave. Thus, action potentials in smooth muscle occur at the peak of pacemaker potentials. Slow waves cannot trigger smooth muscles contraction in the absence of action potentials. Pacemaker potentials are generated by a rhythmic electrogenic sodium-potassium pump whereas action potentials probably represent increased calcium conductance. Slow waves are evident in the distal region of the stomach and in both the small and large intestines. For example, in the human duodenum there

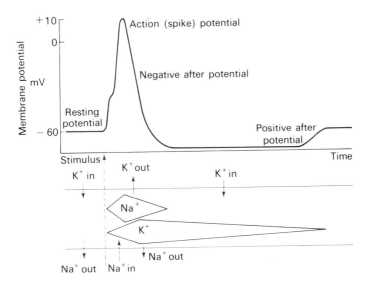

Fig. 1.8   Activation of smooth muscle cell membrane and associated sodium and potassium exchanges (tentative).

are 12 cycles of slow waves each minute, decreasing stepwise to about eight cycles each minute in the distal ileum.

Smooth muscle is also sensitive to stretch and will contract for long periods, unlike striated muscle. When stretched the membrane potential of the cell decreases and the number of spikes increase, initiating contraction. This action may be caused by the release of bound calcium. As spike discharge continues the muscle remains partially contracted, forming *tonic contractions (tonus)*. The number of spikes eventually decrease, the membrane potential is restored and the muscle relaxes.

Gastrointestinal hormones, extrinsic nerves and food may all influence the electrical activity of smooth muscle of the alimentary tract. The hormone gastrin increases and the hormone secretin decreases both the slow waves and action potentials in the stomach. In the duodenum neither hormones affect the slow waves, but gastrin augments and secretin diminishes action potentials in this region. It is still unclear what part hormones play in the control of gastrointestinal motility*. Stimulation of the gastric vagal nerves causes a transient increase in slow wave activity while denervation causes a temporary disorganization of this activity. Feeding increases action potentials in the stomach and duodenum for several hours while temporarily decreasing slow wave activity for a short time.

* Recent evidence suggests that the action of gastrointestinal hormones (gastrin, cholecystokinin-pancreozymin) on visceral smooth muscle activity is mediated by the release of acetylcholine from nerve terminals.

The electrical events in *nerve fibres* are much like those in muscle cells. An action potential is propagated along the nerve fibre, causing a nerve impulse to be carried to the free nerve ending. Small, empty secretory vesicles (diameter 350–450 Å) are seen at the free nerve terminals. Permeability changes occur throughout the length of an unmyelinated fibre, but in a myelinated fibre permeability changes occur only at the nodes which lack the insulating myelin sheath *(saltatory conduction)*. The speed of transmission in myelinated fibres is much faster than that in unmyelinated fibres.

Nerve terminals act like glands secreting a chemical transmitter (p. 9). When a nerve impulse reaches a parasympathetic nerve ending *acetylcholine* is released, which bridges a gap at least 240 Å between nerve and muscle cell, and leads to tonic tension and rhythmic contraction of the muscle. Acetylcholine is associated with depolarization of the muscle membrane, initiates or increases action potentials and causes increased contractions. Acetylcholine is destroyed by

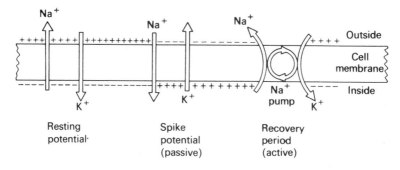

Fig. 1.9   Transfer of sodium and potassium ions across the muscle cell membrane in the resting and active state. It is assumed that ionic shifts for both smooth and striated muscle cells are similar.

the action of an enzyme, *(acetyl)cholinesterase,* at the neuromuscular junction, or it rapidly diffuses away from the receptor site. Little is known about the synthesis and mechanism of storage of (acetyl)cholinesterase.

When a sympathetic nerve is stimulated, *noradrenaline (norepinephrine)* is released at the free nerve ending, leading to the relaxation and loss of tone of the muscle cells. Noradrenaline is associated with the positive after potential of the muscle membrane and reduces or stops action potentials. Enzymes with the properties of (acetyl)cholinesterase have not been identified in the vicinity of sympathetic nerve endings. Movements of the alimentary canal, particularly those of the small intestine, depend more upon the intrinsic action of nerve and muscle cells than upon extrinsic nervous stimulation.

Energy

The energy for muscle contraction comes from the hydrolysis of energy-rich adenosine triphosphate *(ATP)* to adenosine diphosphate *(ADP)*. In the muscle cell ATP is re-synthesized by both ADP and a store of another energy-rich phosphate compound, *phosphocreatine*. At rest, some ATP transfers its phosphate to creatine so that a phosphocreatine store builds up inside the cell. The energy for both ATP and phosphocreatine synthesis in muscle comes from glucose, stored as glycogen. Although the total theoretical yield of glucose is 673 000 calories (1 calorie = 4.1868 joules) per mole, some of this energy is lost as heat. In the cell glucose has to be oxidized before it can release energy (ATP), and is broken down in two stages. First, the glucose molecule goes through a series of anaerobic reactions or *glycolysis* in the cell cytoplasm, yielding pyruvate and two units of ATP (16 000 calories). Secondly, pyruvate is completely oxidized through another series of reactions to yield carbon dioxide and water by the *tricarboxylic acid cycle* and *oxidative phosphorylation* in the mitochondria, yielding 36 units of ATP (288 000 calories). Each series of reactions involves ten enzymes. The net production of ATP is therefore 38 units, that is, the energy recovery-rate (304 000/673 000 × 100) of glucose is about 45 per cent.

2 The salivary glands

Chewing

The teeth are admirably adapted for chewing. The front teeth bite off pieces of food and the back teeth grind them. The combined action of the jaw muscles exerts a considerable crushing force on the food, far in excess of that actually required. Chewing the food makes swallowing easier and increases the surface area of the food to allow digestive enzymes to act on it.

Saliva

Saliva is a slightly acid solution of salts and organic substances secreted by the salivary glands. *Water* accounts for 99 per cent of the secreted fluid and, when mixed with salivary proteins, forms a highly viscid fluid, *mucus*. Man secretes about one and a half litres of saliva each day, a relatively small amount compared with a sheep which secretes about 10 litres a day, and a cow which produces about 150 litres.

The *functions* of the saliva are to lubricate the mouth and moisten the food, to flush out the mouth thus preventing bacterial overgrowth and tooth decay, and to aid in the digestion of carbohydrates. The salivary glands secrete an enzyme, α-*amylase (ptyalin)*, which catalyzes the breakdown of polysaccharides into a mixture of disaccharides. However, our present-day habit of bolting our food hardly gives this enzyme time to act in the mouth, although its digestive action is continued in the stomach for about half-an-hour before being inhibited by acid. Some mammals, for example the cow and sheep, have only a low concentration of this enzyme in their saliva, while others, like cats and dogs, have none. The young calf has another salivary enzyme, pre-gastric *lipase*, that digests milk fats after reaching the stomach.

Structure

There are three pairs of salivary glands, the *parotid, sublingual* and *sub-maxillary,* and also numerous smaller *buccal glands* in the lining of the mouth and soft palate. The salivary glands are composed of various proportions of *serous* and *mucous acini* (cells grouped round a lumen), the first producing α-amylase, salts and water, and the second mucus.

The parotid gland consists mainly of serous acini opening by a single duct on the inside of the cheek opposite the second upper molar tooth. In man the parotid gland produces 25 per cent of the total salivary secretions. The sublingual gland consists mainly of mucous acini, opening by 10 to 20 small ducts on the floor of the mouth beside the tongue, and produces five per cent of the total salivary secretions. Most of the salivary secretions are produced by the submaxillary glands, consisting of mixed acini which open by a single duct on either side of the tongue in the back of the mouth.

Control

The secretion of saliva is controlled entirely by a nervous mechanism. The innervation of the salivary glands was first investigated by C. Ludwig in 1851. The salivary glands are innervated by efferent branches from both the sympathetic and parasympathetic nerves. Both systems stimulate secretion, unlike their antagonistic activity in most other organs.

Figure 2.1 shows the innervation of the salivary glands. A parasympathetic *salivary centre* is situated in a restricted area in the medulla. The anterior portion, the *superior salivary nucleus,* supplies the submaxillary and sublingual glands by a branch (the lingual nerve) of the facial (VII) nerve which synapses in the submaxillary ganglion near the glands. The posterior portion, the *inferior salivary nucleus,* supplies the parotid gland by the glossopharyngeal (IX) nerve, which synapses in the otic ganglion near the gland.

Parasympathetic stimulation liberates an enzyme, *kallikrein*, from the activated secretory cells which acts on plasma protein, kininogen, to form a polypeptide, *bradykinin*, causing the blood vessels surrounding the glands to dilate by relaxing the smooth muscle wall. A copious flow of watery saliva and α-amylase is secreted, that is, the serous acini are stimulated. However, secretion is independent of blood flow for after occluding the blood vessels and then stimulating the parasympathetic nerve, the gland will still secrete.

The sympathetic supply to all three pairs of salivary glands is by the superior cervical ganglion. Short preganglionic fibres originating from the first thoracic segment of the spinal cord synapse in the ganglion and send long postganglionic fibres to innervate the glands. Sympathetic stimulation constricts the blood vessels, when the glands liberate a scanty viscous mucus, that is, the mucous cells are stimulated.

Mechanism

The *tongue* is the organ of taste. Taste is a chemical stimulus, and to detect taste the substance must be in solution. The tongue is covered with stratified squamous epithelium and projecting from its upper surface are three different types of *papillae* (fungiform, foliate and circumvallate). *Taste buds* are embedded within the papillae, which contain *taste receptors*. Although structurally similar, there are four different types of taste buds which respond to sweet, salty, sour or

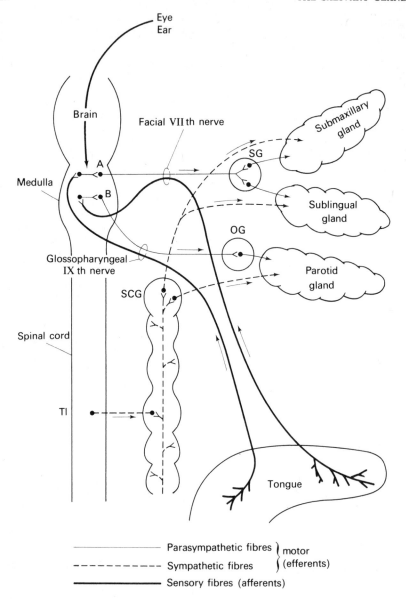

Fig. 2.1   Nervous control of salivary gland secretion. SG: submaxillary ganglion; OG: otic ganglion; SCG: superior cervical ganglion. A: superior salivary nucleus; B: inferior salivary nucleus. The nuclei are located in the salivary centre in the medulla. Details in text.

bitter taste (Fig. 2.2). The sensation for a sweet or salty taste is detected by the tip of the tongue, that for a sour taste by the sides of the tongue, and that for a bitter taste by the back of the tongue.

Substances in solution enter the taste receptors (chemoreceptors) and stimulate gustatory cells there. Nerve impulses are then relayed to the salivary centre in the medulla by sensory (afferent) fibres of the mixed lingual (facial) nerve, which innervates the anterior two-thirds of the tongue, and the mixed glossopharyngeal nerve which innervates the posterior one-third. The salivary centre then integrates and controls the salivary secretions by the autonomic nervous system (Fig. 2.1).

The parasympathetic nerves are more important than the sympathetic nerves, since denervation of the parasympathetics but not the sympathetics abolishes the

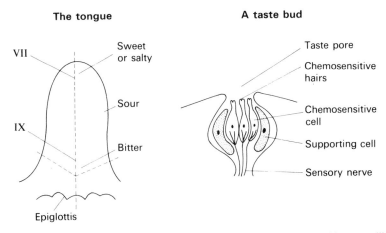

Fig. 2.2   Taste areas of the tongue and structure of a taste bud in a papilla.

normal reflex action. The role of the sympathetic nerves is obscure. The kind of salivary reflex causing increased secretion at meal-times is called an *unconditioned (inborn) reflex*.

There exists, however, another kind of salivary control which is termed a *conditioned reflex* and depends on experience. It is controlled by sight and thought of food. The banging of pots and pans in the kitchen will also cause a hungry individual to salivate in anticipation of a satisfying meal. Sight of food becomes associated with the taste of food, and ingoing afferent nervous stimuli are sent by the eyes and tongue to both visual and salivary centres in the brain. Association pathways become established between these two centres and impulses are then transmitted by the autonomic nervous system directly to the salivary glands. Provided that the stimulus looks delectable, the sight of food alone may evoke salivary secretion through the eyes, brain and salivary glands.

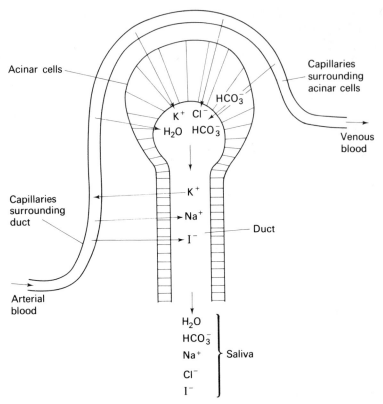

Fig. 2.3 Electrolytic exchanges during salivary secretion (after Texter *et al.*, 1968).

Electrolytes

The secretions of the salivary glands also include various electrolytes. Figure 2.3 illustrates the electrolytic exchanges with the blood during the secretion of saliva. Water, potassium and chloride ions are derived from the blood capillaries surrounding the acini and bicarbonate is derived from both the blood and the acini. Potassium ions are mainly re-absorbed from the fluid in the salivary ducts into the capillaries surrounding the ducts in exchange for sodium ions. Since potassium ions are re-absorbed more rapidly than sodium ions are delivered, the saliva secreted is hypotonic. When the salivary rate is increased sodium and chloride levels also increase although their concentration remains lower than in the blood. In addition, the salivary glands extract most of the iodine from the blood and concentrates it in the saliva. Ruminants also secrete substantial amounts of urea in their saliva.

3 The oesophagus

Structure

The oesophagus is a muscular tube extending from the mouth and pharynx to the stomach. It acts as a passageway for food and liquid. Neither digestion nor absorption takes place there. The oesophagus is lubricated by mucus from the salivary, nasal and bronchial glands and, in man, by mucus from glands in the submucosa. Sphincters are found at either end of the oesophagus. The *upper oesophageal sphincter* consists of a thickened band of muscle. The *lower oesophageal sphincter* is non-muscular but otherwise has a similar structure to that of the oesophagus. Both the sphincters remain tonically contracted when swallowing does not take place, thus closing both exits of the oesophagus.

The oesophagus is lined by non-keratinized squamous epithelium, several layers thick, typical of a 'wear and tear' surface. The basal layer of cells proliferate and continually push non-proliferating cells above the basal layer into the lumen, a process taking three to five days. The muscle coat consists of an inner circular and outer longitudinal layer. In man, the upper one-third of the oesophagus consists of striated muscle, the lower one-third of smooth muscle, and the middle region of mixed striated and smooth muscle. Most other mammals have striated muscle throughout the oesophagus.

The entire oesophageal region, including the sphincters, is innervated by parasympathetic nerves of the vagus. The sympathetic innervation is by fibres from the superior cervical ganglion (the upper sphincter) and the coeliac ganglion (the oesophagus and lower sphincter). Oxygenated blood from the abdominal aorta is supplied to the oesophagus by a branch of the coeliac artery and deoxygenated blood removed to the liver by a branch of the portal vein.

Swallowing

A gulp of beer trickles down the oesophagus by gravity, unless you drink it standing on your head, in which case the liquid will be handled like a solid ball. The act of swallowing a mouthful of crisps is, however, surprisingly complicated.

The chewed food is first squeezed and rolled into a solid ball or *bolus* and forced to the rear of the mouth into the pharynx by successively lifting the front and then the rear of the tongue. To avoid the food going down 'the wrong way' the nasal cavity is sealed by the soft palate and the glottis by the epiglottis, so preventing the food from reaching the trachea. As a result of these various muscular activities the pressure in the pharynx rises abruptly.

Figure 3.1 summarizes the sequence of pressure changes during a swallow. The movement of the food bolus along the oesophagus depends upon a series of peristaltic waves of the oesophageal wall pushing the food downwards. The upper oesophageal sphincter muscles, which are normally contracted, relax momentarily, opening the lumen and allowing the bolus to shoot into the oesophagus. The sphincter then closes again. A *negative pressure wave* in the lumen is first formed, which rapidly travels down to the base of the oesophagus and relaxes the lower oesophageal sphincter. A slower *primary positive pressure wave* then follows, sweeping the food bolus ahead of the pressure gradient. when the bolus reaches the relaxed lower sphincter it passes unhindered into the stomach. The lower sphincter contracts again and the pressure in the oesophageal lumen rises slightly above ambient pressure, preventing food in the stomach from re-entering the oesophagus. How this is caused is unknown. Primary propulsive waves are followed by *secondary positive pressure waves,* resulting from previous distension of the oesophagus wall. These waves re-inforce the propulsion of the bolus along the oesophagus.

Control

The act of swallowing is mainly a *somatic motor process.* When the food bolus reaches the pharynx, impulses from the pressure receptors in the walls send

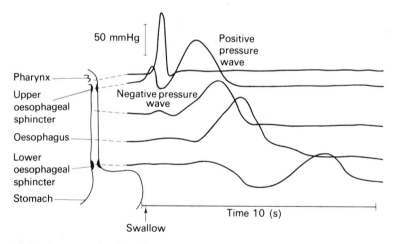

Fig. 3.1   Pressure in the pharynx and oesophagus during swallowing. The pressure in the pharynx rises sharply. The upper oesophageal sphincter relaxes briefly. A negative pressure gradient quickly travels down the oesophagus and relaxes the lower oesophageal sphincter. A slower positive peristaltic pressure wave then pushes the food bolus down the oesophagus, which empties into the stomach through the relaxed lower oesophageal sphincter (modified from various authors).

afferent (sensory) impulses mainly by the vagus to the *swallowing (deglutation) centre* in the medulla, which co-ordinates the swallowing process by the muscles of the pharynx and oesophagus. Efferent (motor) impulses from the swallowing centre via the vagus briefly relax the upper oesophageal sphincter, allowing the food into the oesophagus. Afferent stimuli from the thoracic branches of the vagal nerve also initiate a reflex contraction of the oesophageal muscles above the bolus, and the food is pushed downwards. As the bolus nears the lower oesophageal sphincter it opens by a reflex mechanism similar to that of the upper oesophageal sphincter, and the food empties into the stomach.

The upper and lower sphincters are normally constricted owing to impulses from sympathetic fibres from the superior cervical ganglion and coeliac ganglion respectively. Several gastrointestinal hormones (gastrin, secretin, cholecysto-kinin-pancreozymin) are also able to regulate the tension in the lower sphincter. Sympathetic stimulation leaves peristaltic activity unaltered.

4 The stomach

Most mammals have a single-chambered or *monogastric* stomach. Typically, ruminants have a multi-chambered or *digastric* stomach.

The monogastric stomach

Structure

The stomach is a dilatable sac which, in man, holds two and a half litres. The stomach wall is divided into the typical mucosal, submucosal, muscular and serous layers. The mucosa has a corrugated surface *(rugae)* and the epithelium is invaginated to form numerous glands. An inner, incomplete, oblique muscle layer overlies the circular and longitudinal muscle layers. The circular muscle layer gradually thickens in the lower portion of the stomach.

The stomach is divided into the *cardia, fundus* (or body) and *pyloric antrum,* each containing different types of glands (Fig. 4.1). The relative proportions of these different regions vary in different species (Fig. 4.2). The inner surface of the stomach is studded by openings of the *gastric pits.* Each gastric pit leads to several branched tubules arranged perpendicularly to the surface, the *gastric glands,* which are embedded in the underlying connective tissue.

The cardiac portion of the stomach lies near the oesophageal opening and con-tains branching *cardiac glands* that secrete mucus, electrolytes and some hydrochloric acid. The fundic portion is the central and largest region, and con-tains tubular *fundic* or *gastric glands* which are composed of the following three main types of cells:

1 *Peptic* (chief or zymogenic) *cells,* which are found lining the base of the gland. These produce pepsinogen which is activated by hydrochloric acid to pepsin. They may also provide intrinsic factor. They are the most abundant of the cell-types.
2 *Oxyntic* (parietal) *cells* are found mainly in the upper region of the gland and secrete hydrochloric acid and water. Depending on the species, these cells may also produce intrinsic factor.
3 *Mucoid cells,* which are found mainly along the sides of the gastric pits and secrete mucus. Occasional enterochromaffin cells may also be seen.

The *pyloric antrum (pylorus)* lies above the pyloric sphincter and contains

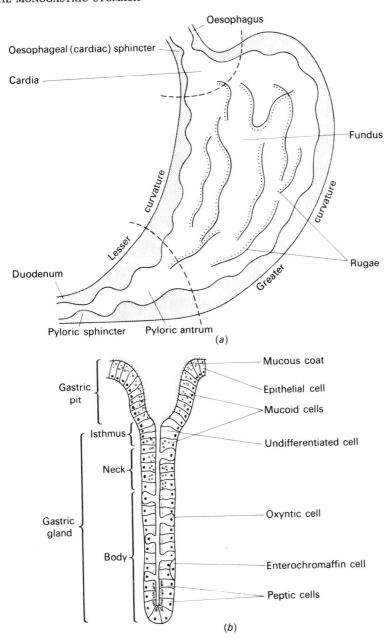

Fig. 4.1 Structure of the monogastric stomach. (a) The main physiological divisions. (b) Detail of a fundic (gastric) gland. The peptic cells secrete pepsinogen and the oxyntic cells secrete hydrochloric acid (modified from Passmore and Robson, 1968).

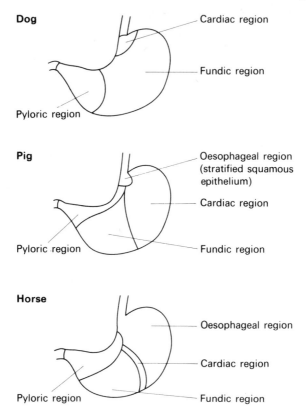

Fig. 4.2   Variations in the surface area of the main regions of the monogastric stomach in different mammals (adapted from various authors).

branching *pyloric glands* that produce mucus, some hydrochloric acid and a hormone, *gastrin*, which controls the release of hydrochloric acid from the oxyntic cells. Gastrin is synthesized in granular enterochromaffin-like cells (endocrine G-cells), which are dispersed in the upper half of the pyloric glands. At least four gastrins, differing in molecular weights, euphemistically called, 'mini', 'little', 'big' and 'big, big', have recently been isolated. In the human stomach mucosa at least four endocrine cell types have been discovered but only one, gastrin G-cells, has been associated with a known hormone.

Columnar *epithelial cells* line the surface of the gastric mucosa and openings of the gastric pits. They are formed by mitoses from *undifferentiated cells* lining the base of the gastric pits (the isthmus), and are pushed to the gland opening and shed into the stomach lumen one to three days after being formed. Peptic and oxyntic cells, however, survive much longer.

Both ends of the stomach are normally closed by sphincters. This is an unfortunate term since neither of them are true sphincters. The lumen of the oesophagus and stomach junction is partially obstructed by a overhanging mucosal valve-like flap, the *cardiac (lower oesophageal) sphincter*. The distal end of the pyloric antrum tapers to form a narrow open funnel, the *pyloric sphincter*, leading to the entrance of the small intestine.

The stomach is innervated by two branches of the vagus that synapse in ganglia in between the circular and longitudinal muscle layers and in the submucosa, from which fibres penetrate the mucosa. Sympathetic fibres synapse in the superior mesenteric and coeliac ganglia and then send fibres to innervate the stomach wall. Sympathetic stimulation stops stomach movements and secretory activities; parasympathetic stimulation initiates movements of the stomach and its secretions. If the vagi are cut gastric motility stops but eventually returns owing to the activities of the intrinsic nerve plexus. Oxygenated blood is supplied by three branches of the coeliac artery; deoxygenated blood is removed to the liver by three branches of the portal vein.

Secretions

The stomach acts as a reservoir for ingested food which is partially digested there by the action of gastric secretions. The food is converted into a semi-solid mass in a corrosive liquid known as *chyme*. The gastric juice contains hydrochloric acid, a proteolytic enzyme, pepsin, intrinsic factor, electrolytes and mucus. The stomach takes four to six hours to empty.

Hydrochloric acid is secreted by the oxyntic cells following vagal stimulation. How they do this is still unsolved, although various mechanisms have been proposed (Fig. 4.3). Man secretes about two litres of hydrochloric acid each day. This bathes the chyme as a 0.5 per cent acid solution (pH 1.5–2.0). Hydrochloric acid helps to break up the molecules of protein bonds, activates some enzymes secreted by the stomach, and kills bacteria in the lumen entering with the food. The release of hydrochloric acid into the lumen is accompanied by discharge of bicarbonate into the bloodstream.

The main enzyme in the stomach is *pepsin,* which is secreted by the peptic cells following stimulation by the vagus, intrinsic nerves and the hormone secretin. It is secreted in an inert form (or proenzyme), pepsinogen, which is converted into the active form, pepsin, in an acid medium by splitting off part of the pepsinogen molecule (several types of pepsinogens have been isolated). Pepsin is a proteolytic endopeptidase (proteinase), attacking inner protein linkages adjacent to free carboxylic groups, and forms large peptide molecules.

Another proteolytic endopeptidase, *rennin,* has been isolated from the calf stomach. The location of rennin secreting cells is unknown. This enzyme, which works in a sub-optimal pH (optimal pH 5.5), clots milk by converting caseinogen

Blood   Oxyntic cell   Lumen of stomach

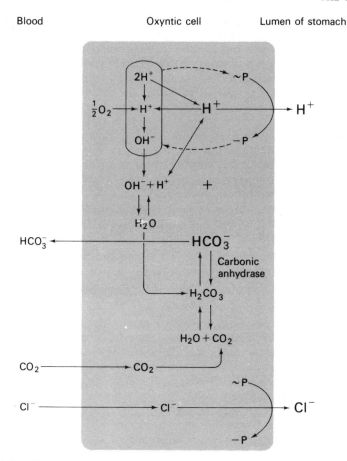

Fig. 4.3 Hypothetical scheme for the intracellular control of pH in the gastric mucosa. Intracellular H^+ ions are available both from the dissociation of carbonic acid ($H_2+CO_2 \rightleftarrows H_2CO_3 \rightleftarrows H^++HCO_3^-$) and from oxidative phosphorylation in the mitochondria (top left). H^+ is secreted into the lumen of the stomach by an energy-dependent process, requiring ATP generated by the mitochondria. HCO_3^- diffuses into the blood while Cl^- from the blood is actively secreted from the oxyntic cell into the lumen accompanying the active secretion of H^+ to form HCl (modified from Hersey, S. J. (1971) *Phil. Trans. Roy. Soc. Lond.* B. **262**: 272).

to casein in the presence of calcium ions, which then precipitates as calcium caseinate. The level of rennin gradually falls from birth and the level of pepsin gradually rises. Rennin is missing in man, its function being performed by chymotrypsin (p. 59). There is some evidence to suggest that rennin is secreted in infants when the stomach contents are less acid than in adults. A weak gastric *lipase,* of little significance, has been isolated which may also be important in infants.

Gastric *mucus* contains various mucopolysaccharides and mucoproteins. Mucus prevents the stomach wall from digesting itself by forming two layers, one over the surface of the gastric mucosa, and another in between the mucosal cells. Although hydrochloric acid can penetrate these barriers, it is prevented from damaging the cell cytoplasm by becoming buffered by alkaline electrolytes trapped within the mucous layer. Mucus also tends to inactivate pepsin. Glandular mucoproteins contain *intrinsic factor* which aids the absorption of vitamin B_{12}.

Control

When food enters the stomach, gastric juice secretion begins and continues for about four hours. Gastric activity has been studied in detail in dogs, in which a portion of the main body of the stomach was surgically formed into an isolated stomach pouch or miniature stomach with an attached fistula through which secretions can be measured and so indicate what is happening in the rest of the stomach (Fig. 4.4). Such a pouch can have an intact nerve supply (Pavlov pouch) or both the external and internal nerve connections can be separated, leaving the pouch denervated (Heidenhain pouch). In both types of pouches, the blood supply to the stomach is left intact. Gastric secretion can also be studied without food reaching the stomach by sectioning the oesophagus and suturing the cut ends to the skin surface in the neck (oesophagostomy). Gastric secretions through a gastric fistula can then be measured after 'sham feeding' the animal. When the animal eats, the food bolus emerges from the neck and is rejected externally.

When a dog is given a meal and gastric secretions collected from an innervated (Pavlov) pouch, a bi-phasic secretory response is recorded. Gastric secretions reach a peak after about half an hour, then decrease, and another lower peak occurs after one and a half to two hours. Secretions stops altogether about four hours from the start of the meal. The results of further experiments (described below) proved that gastric secretion is stimulated by at least two mechanisms, involving a nervous reflex and hormonal action.

After 'sham feeding' an oesophagostomized dog gastric secretion begins soon after the food is taken, reaches a peak in about half an hour, and ends after about one hour. The gastric juice is highly acid and rich in pepsin (Fig. 4.5*a*). The initial stimulus of gastric activity *(cephalic phase)* is due to both conditioned and unconditioned reflexes. Like salivary secretion, the sight and smell of food evoke gastric secretion by conditioned reflexes. On eating, the food stimulates receptors in the buccal cavity lining which initiate unconditioned reflex gastric secretion following stimulation by the *vagal nerve.* Gastric juice is not secreted if the vagus is cut.

The denervated (Heidenhain) stomach pouch, which does not secrete during 'sham feeding', secretes both acid and pepsin when food enters the stomach *(gastric phase).* But the rate of secretion is slower, reaching a peak after one and a

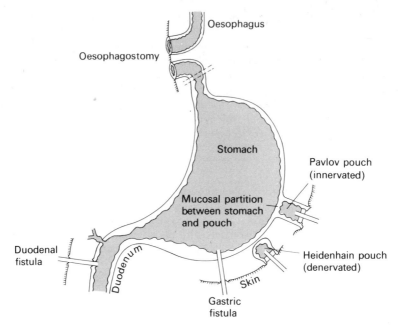

Fig. 4.4   Pouches associated with the study of digestive secretions in the mono-gastric animal. Such pouches enable gastric juice to be collected in the absence of food. The oesophagus can also be divided and both ends fixed to the surface of the animal, which is then 'sham fed' (details in text).

half to two hours and ending after about four hours. Since nerves cannot stimulate the production of gastric juice, the stimulus is not a reflex action but must be mediated by some other mechanism.

The only pathway connecting the denervated pouch to the rest of the stomach is the circulation, indicating that a hormone might stimulate secretion. When the hormone *gastrin* is injected into the blood of a normal dog, the stomach secretes over a period of several hours (Fig. 4.5*b*). When food, particularly proteins, enters the stomach and distends its wall, gastrin is reflexly released by endocrine 'G' cells in the glands of the pyloric antrum and absorbed into the bloodstream where it reaches the fundus and stimulates the secretion of a highly acid juice with some pepsin. Small amounts of gastrin are also produced in the duodenal mucosa and in pancreatic islet tissue. When the acidity of the gastric contents drops to a pH of 3.5 the release of gastrin from the pylorus begins to decrease, and when the pH reaches 1.5 the release of gastrin stops. The mechanism by which acid affects gastrin release is unknown. In addition, the secretion of *histamine* by the gastric mucosa releases a small volume of hydrochloric acid

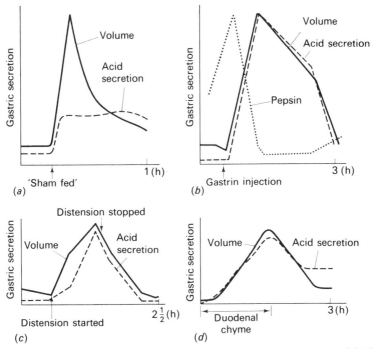

Fig. 4.5 Control of gastric secretion in the dog. Secretion in response to (a) 'sham feeding', (b) a subcutaneous injection of gastrin, (c) distending the stomach and (d) irrigating the stomach with duodenal food (after Gregory, 1962, modified from various authors).

from the fundic region. Finally, mechanical *distension* of the stomach by food will also cause a small production of hydrochloric acid by a local reflex action (Fig. 4.5c).

Figure 4.6 summarizes the different mechanisms of gastric secretion in a monogastric animal. The summation of the secretion produced by these mechanisms gives the total gastric response to a meal.

The effects of placing food directly into the duodenum through a duodenal fistula, or irrigating the stomach with food taken from the duodenum, have also been studied (Fig. 4.5d). When this is done, a slow flow of gastric secretion starts after a time-lag and continues for some time *(intestinal phase)*. Since a denervated pouch also secretes in this way the stimulus is probably a gastrin-like hormone ('intestinal humoral agent'), but the hormone has not yet been isolated.

There are several hormones that stop the secretion of gastric juice when a sufficient amount has been produced to digest the food, thus preventing its corrosive action on the lining of the stomach and duodenum. Accumulation of acid in the stomach inhibits the production of gastrin and hence acid (above). In addition, the passage of acid food in the proximal small intestine stimulates the release of *secretin* and *bulbogastrone*. Partially digested proteins release *cholecystokinin-pancreozymin*, and fats release *enterogastrone*(s) (the existence of this hormone is controversial). *'Gastric inhibitory polypeptide'* and *'vasoactive intestinal polypeptide'* have also been implicated. These hormones are all synthesized in the mucosa of the proximal small intestine, then liberated into the bloodstream and reach the stomach wall to inhibit gastric secretion and motility (Fig. 5.1). Finally, the presence of hypertonic solutions in the duodenum may also have an inhibitory effect on gastric secretion.

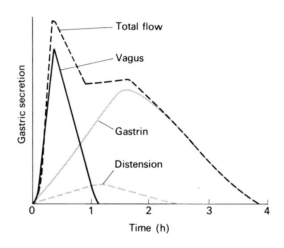

Fig. 4.6   The different phases of gastric secretion.

Movements

When empty, the stomach is small and relaxed. Two main types of peristaltic contractions are evident, originating in the upper half of the stomach (Fig. 4.7). Slight ripples of contractions (Type I waves) pass from the cardia towards the pylorus. These movements are not propulsive but mix the food, and depend on the integrity of the intrinsic nerve plexus. Occasionally, larger peristaltic waves (Type II waves) sweep over the stomach owing to increased vagal motor activity.

As food enters the stomach the upper region relaxes to accommodate the food and pronounced gastric contractions then follow in response to distension of the stomach wall by the food. Stretch receptors in the mucosa increase afferent vagal impulses which initiate propulsive waves and vigorous peristaltic contractions, originating from the muscle layers in the upper region of the stomach, sweep and intensify towards the pyloric sphincter at the rate of about three cycles each minute. These waves mix the food and push it towards the pyloric antrum. Chyme is squirted through the pyloric sphincter mainly by repeated contractions of the antral muscles. The pyloric sphincter functions as a narrow passageway to prevent regurgitation of intestinal contents into the stomach. A hormone, *motilin*, has been identified in enterochromaffin (EC) cells in the duodenal mucosa.

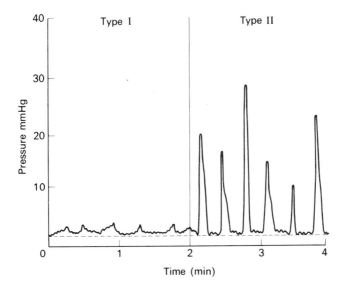

Fig. 4.7 Main peristaltic waves in the pyloric antrum of the empty monogastric stomach. Type I (mixing) waves have a low amplitude and type II (propulsive) waves have a greater amplitude, each lasting for about 20 s (modified from Texter *et al.*, 1968).

This hormone is produced in response to alkaline contents in the proximal intestinal lumen and thought to be partly responsible for stimulating stomach movements.

The stomach starts to empty soon after it is filled. A solid meal takes at least four hours to leave the stomach. The rate of gastric emptying depends on the interplay of several factors, such as the amount and physical nature of the food in the stomach and the presence of hypertonic solutions, by controlling mechanisms not yet fully understood.

Both the quantity and quality of food reaching the duodenal lumen stimulate stretch receptors in the duodenal mucosa which reflexly inhibit stomach movements. Also, various hormones synthesized by endocrine cells in the proximal small intestinal epithelium decrease gastric motility (above and Table 7.2). During emotional stress both movements and secretory activities of the stomach and small intestine stop, resulting from sympathetic inhibition.

Absorption

Solid food and water remain virtually unabsorbed in the stomach, but alcohol is absorbed. A glass of milk before a stag party slows down both the rate of alcohol absorption and gastric motility. Unfortunately, alcohol is eventually absorbed by the small intestine, so a hangover will be delayed, not prevented. However, aspirin (acetylsalicylic acid) is rapidly absorbed by the stomach; a comforting thought.

The digastric stomach

Structure and functions

The digastric, or multi-chambered, stomach is divided into four chambers: the rumen, reticulum (or reticulo-rumen; that of the sheep originally formed the wrapping of the Scots' genuine haggis), omasum and abomasum (Fig. 4.8). The first three compartments are sometimes called the 'forestomach' and the last compartment the 'true stomach'. The capacity of the stomach in relation to other regions of the alimentary canal in monogastric or digastric animals differs widely (Table 4.1). In the pig the monogastric stomach occupies about one-third of the total volume of the alimentary canal, while in the ox or sheep the digastric stomach occupies over two-thirds, three-quarters of which is rumen. The combined large intestine and rectum in the horse and pig occupy one-half and one-third of the total capacity respectively, whereas in the ox or sheep they occupy about one-tenth. In farm animals the small intestine occupies between one-fifth and one-third of the total volume of the alimentary canal.

THE RUMEN In the adult the rumen occupies by far the largest capacity of the digastric stomach. In the cow the rumen has an average capacity of about 100 litres and in the sheep 10 litres. The rumen is a large bag asymmetrically placed in the left side of the abdominal cavity, lying behind the reticulum. The rumen is partly sub-divided into four chambers by muscular invaginations *(pillars)* of the walls: a large *dorsal* and *ventral sac* and smaller *dorsal* and *ventral blind sacs*. The entire surface area is increased by prominent *papillae,* consisting of stratified squamous epithelial cells, the surface cells of which are continually being slouched into the lumen following cell division in the basal layer. The

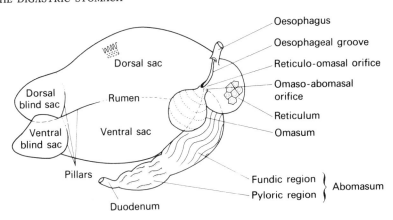

Fig. 4.8   The digastric or multi-chambered stomach of an adult ruminant.

papillal in the dorsal sac are larger than those in the ventral sac. The number and size of the papillae depend upon the type of food eaten, which in turn depends more upon the chemical nature of the food than upon its physical condition. Thus, the papillae remain few and small when a young animal is continually fed on milk and become large and numerous if it is then fed on a mixture of milk and hay. Although the rumen, reticulum and omasum lack mucosal glands the main regions generally have the same anatomical arrangement as that of the monogastric stomach, including the nervous and vascular systems.

In the young ruminant, the milk suckled by-passes the reticulo-rumen and goes directly into the abomasum through the *oesophageal groove* (the omasum develops later). During suckling the groove closes, partly as a result of a vagal reflex from efferent stimuli in the mouth and pharynx, forming a channel between oesophagus and abomasum. The presence of sodium chloride and bicarbonate ions in the food and certain behavioural responses also stimulate closure.

Table 4.1   Total capacity of different regions of the alimentary canal in digastric and monogastric animals (adapted from Phillipson in Swenson, 1970)

| | Percent total capacity | | | |
| | Digastric | | Monogastric | |
Alimentary canal	Ox	Sheep	Horse	Pig
Whole stomach	71	67	9	29
Reticulo-rumen	60	57	—	—
Omasum	5	2	—	—
Abomasum	6	8	—	—
Small intestine	18	21	30	33
Caecum	3	2	16	6
Large intestine and rectum	8	10	45	32

As the young ruminant grows the relative proportions of the different chambers of the stomach change markedly (Table 4.2). Thus, in the new-born digastric the reticulo-rumen occupies about one-third of the total volume of the stomach while the abomasum occupies about three-quarters. In the adult, the reticulo-rumen occupies about four-fifths of the total capacity and the omasum and abomasum occupy the remainder.

Unlike the monogastric stomach, the digastric stomach both digests and absorbs food. Over three-quarters of the total food intake is digested in the rumen. The rumen is a large fermentation incubator partly filled with a smelly vegetable soup in which a vast population of symbiotic *bacteria* and ciliated *protozoa* digest the food. One cm^3 of rumen fluid contains on average 10^{10} bacteria and 10^6 protozoa, or about four per cent of the total volume of the fluid. There is a pocket of anaerobic gas above the fermenting liquid, containing mainly carbon dioxide (65 per cent) and methane (27 per cent), and small amounts of nitrogen, oxygen, hydrogen and hydrogen sulphide. Methane is formed from reduction of carbon dioxide. The temperature of the fluid averages 40°C and the pH 6–7, which is buffered by an influx of large amounts of saliva containing bicarbonate and phosphate.

Table 4.2 Relative percent volume of different chambers of a ruminant at various ages (modified from various authors)

Age	Reticulo-rumen	Omasum	Abomasum
Newly born	30	0	70
Two months	50	0	50
Three months	70	0	30
Four months	80	0	20
Eighteen months (adult)	85	7.5	7.5

The products of digestion are mainly a mixture of short-chain fatty acids, *volatile fatty acids* (VFA) or *steam fatty acids,* three-quarters of which are absorbed by the rumen and reticulum and most of the remainder by the omasum and abomasum. Some of the VFA are utilized by bacteria and protozoa themselves. Typically, VFA consist of 65 per cent acetic, 25 per cent propionic and 10 per cent butyric acids. However, the food intake, type and physical condition of the diet all affect the chemical composition of VFA. Thus, on silage, the proportion of acetic acid will rise, on maize the proportion of propionic acid will rise, and on hay the proportion of butyric acid will rise. Sodium ions are also absorbed in the reticulo-rumen while chloride and inorganic phosphate move in either direction. In some herbivores both the caecum and large intestine are enlarged to accommodate a varied population of bacteria and protozoa, which ferment VFA from undigested cellulose which can then be absorbed.

Excess gases produced by fermentation periodically escape from the rumen by *eructation*. Movements of the wall of the sac from a posterior to anterior direction push the digesta forwards and downwards, allowing the gases to escape through the oesophageal opening.

The rumen of a newly born animal is empty but soon acquires a population of bacteria (mainly lactobacilli) from milk regurgitated from the abomasum. At first the rumen lacks a protozoan population, but acquires one after two to three weeks when the animal begins to feed on grass. As the diet changes, so the bacterial population alters. When the animal matures, the bacterial and protozoan population stabilize. About 30 genera and over 60 species of rumen bacteria have been isolated from rumen fluid, and about 15 genera and 40 species of protozoa identified. Marked fluctuation in numbers and species composition may occur when the internal environment changes, depending on the state, type, and way in which the food is given.

THE RETICULUM The reticulum forms a continuation of the rumen and, as its name suggests, it has a reticulated appearance of honeycomb-like ridges which are covered with small papillae. Its digestive function is like that of the rumen. The reticulo-rumen opens into the omasum by the reticulo-omasal orifice.

THE OMASUM The omasum is a small, round, compact structure, the inner surface area of which is increased by many thin folds, or *laminae,* of various lengths. The omasum both filters and mixes food particles, and absorbs water and VFA. The omasal-abomasal orifice lies close to the reticulo-omasal orifice.

THE ABOMASUM This is an elongated chamber divided into two main areas. The orifice from the omasum leads into the proximal region *(fundic area)* which is lined by a glandular mucous membrane folded into several long spiral folds. This region produces hydrochloric acid, pepsin, some gastrin and mucus, and in the young calf, rennin. The distal region *(pyloric area)* has a corrugated inner surface, like the monogastric stomach, is more muscular and contains numerous mucus and gastrin producing glands. The fluid entering the abomasum has a pH of 6 and that leaving it has a pH of 2.5.

Digestion and absorption

(a) CARBOHYDRATES The fermentation rate of different carbohydrates varies. Soluble sugars are rapidly fermented, starches less rapidly, and cellulose and hemicellulose only slowly fermented. Some cellulose escapes digestion in the stomach to be digested later in the caecum and large intestine.

Carbohydrates constitute about one-quarter of the total dry matter of summer grass. The dry matter has the following composition: cellulose 10–35 per cent, hemicellulose 10–25 per cent, lignin 2–12 per cent and soluble carbohydrates (hexoses, sucrose and fructosan) about 25 per cent. The grass is fermented by bacteria and certain protozoa into a mixture of VFA and anaerobic gases.

Most of the acetic acid is absorbed through the rumen epithelium into the portal blood to be metabolized by the tricarboxylic acid cycle in the liver. Some of the propionic acid is first converted into lactic acid in the rumen epithelium and then metabolized by the tricarboxylic acid cycle in the liver; the remainder is converted into blood glucose. Most of the butyric acid is first converted into ketone bodies in the rumen epithelium to eventually enter the tricarboxylic acid cycle in the liver. VFA provide an important source of energy in the form of ATP to support the growth of bacteria and protozoa, which in turn provide a notable source of protein for the ruminant. This is particularly important in the ruminant animal since glucose, a major energy end-product in the diet of the non-ruminant, is virtually missing in the ruminant diet. VFA also affect the chemical composition and yield of milk in ruminants. Generally, acetic acid increases milk yield and its constituents, while both propionic and butyric acids affect the composition of milk constituents.

(b) PROTEINS Figure 4.9 summarizes the digestion of proteins in the ruminant alimentary tract. The nitrogen content of summer grass accounts for 15–25 per cent of its total dry matter, over three-quarters of which is protein nitrogen and the remainder non-protein nitrogen. Proteins are digested by proteolytic enzymes in the rumen which are provided by both rumen bacteria and protozoa. The main end-product of protein digestion is *ammonia*. Ammonia fermentation proceeds by hydrolysis through peptides and amides to amino acids. The amino acids are largely destroyed by fermentation and deaminated to ammonia, some VFA and carbon dioxide. The digestion of nucleoproteins also add to the ammonia supply in the rumen.

Some of the ammonia is absorbed directly into the bloodstream to be resynthesized to amino acids or converted to urea. Some is resynthesized to bacterial proteins. Since bacteria are themselves later digested in the abomasum and small intestine, bacterial proteins are converted by the ruminant's enzymes into peptides and amino acids which are absorbed and incorporated in the body tissues.

In ruminants *urea*, the major end-product of nitrogen excretion, is not all excreted. Some of it is returned by the bloodstream to the rumen as urea in the saliva. Urea is converted rapidly by the enzyme urease in the rumen epithelial cells and by bacteria to ammonia and carbon dioxide and then released in the lumen or incorporated into bacterial proteins. Undigested proteins in the rumen are later digested in the abomasum and small intestine.

(c) FATS Pasture grass contains lipids amounting to only about 10 per cent of the total dry matter. The lipids include mainly triglycerides and long-chain unsaturated fatty acids, with smaller amounts of waxes, sterols and phospholipids. Dietary triglycerides are directly hydrolyzed by rumen bacteria and some species

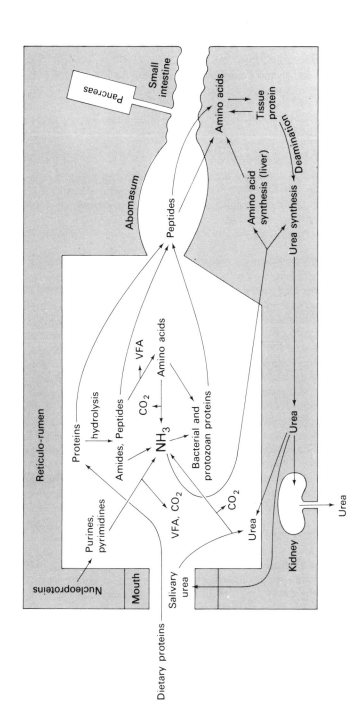

Fig. 4.9 Digestion of proteins and protein regeneration in the ruminant (modified from Haupt in Phillipson, 1970).

of protozoa to glycerol and free fatty acids. In contrast to non-ruminants, the intermediates diglyceride and monoglyceride have not been detected in the rumen. Unsaturated fatty acids (mainly linolenic acid) are fermented by bacteria to saturated fatty acids (mainly palmitic and stearic acids) by hydrogenation and passed to the abomasum and, small intestine to be absorbed. Bacteria can also synthesize long-chain fatty acids from VFA.

(d) VITAMINS The bacteria in the rumen are capable of synthesizing water-soluble B complex vitamins. Dietary cobalt is required for this synthesis. Vitamin B_{12} is absorbed mainly in the distal small intestine while other B complex vitamins are absorbed mainly proximally. Except for vitamin K other fat-soluble vitamins are not synthesized.

Movements

Movements of the digastric stomach are very complex and involve a series of contractions that result in pressure changes in the lumen (Fig. 4.10). First the reticulum contracts, then relaxes and contracts again *(biphasic contraction)*. Before the second reticular contraction is complete a vigorous wave of backward-moving contraction *(primary wave)* begins flowing along the dorsal sac of the rumen, to be followed by a similar wave flowing along the ventral sac. Then a wave of contraction *(secondary wave)* sweeps forwards, first over the dorsal sac and then over the ventral sac. As a result the ingesta is forced upward into the anterior region of the dorsal sac. Successive pressure waves take place at one-minute intervals or less. Contractions decrease when the animal is resting and increase when it is either eating or ruminating.

Successive cycles of contractions circulate the semi-solid digesta within the reticulo-rumen. The coarse particles are gradually compressed into the dorsal sac, leaving a pocket of anaerobic gas above the digesta while finer food particles and fluid sink into the ventral sac and reticulum. Less dense food particles floating on the fluid are regurgitated, chewed and swallowed again, to be further fermented and eroded in the ventral sac and reticulum.

For most of the reticulo-rumen cycle the reticulo-omasal orifice remains loosely open. During the second reticular contraction the orifice opens widely, and as semi-liquid food enters the omasum the pressure in the omasum and reticulo-omasal orifice falls abruptly. The orifice then closes, the pressure in the omasum rises again and forces liquid material through the omasal-abomasal orifice into the abomasum. The passage of food into the omasum continues throughout the day, controlled by the valve-like action of the reticulo-omasal orifice. Solid food remains in the reticulo-rumen and omasum of the cow for 60 to 80 hours, depending on the nature of the diet, and 20 to 40 hours in the sheep.

In contrast to the other chambers, food flows slowly and continually through the abomasum and is squirted into the duodenum in short gushes. The fundic region is relatively inactive and the liquid food there is first mixed with a copious,

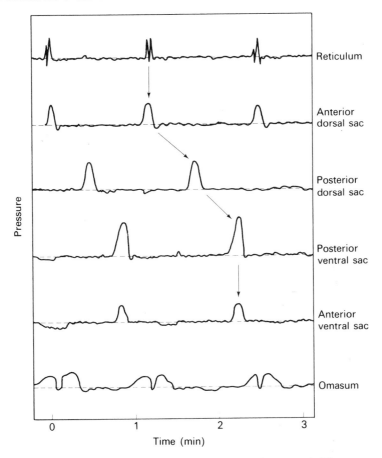

Fig. 4.10   Records of pressure changes in the digastric stomach. The progress of one cycle is indicated by arrows (adapted from various authors).

highly acid juice containing mainly pepsin and, in the young calf, rennin. Bacteria and protozoa entering the abomasum are killed by the acid. Contraction waves sweeping over the pyloric region then mix the food with a neutral mucus. In between meals about 250 cm^3 of liquid passes through the abomasum of the calf, which increases to about 800 cm^3 after feeding; in the sheep it rises from about 200 cm^3 to 500 cm^3.

Rumination is associated with an extra reticular contraction *(triphasic contraction)* preceding the normal double (biphasic) contraction. As contraction starts, the glottis closes and the pressure in the reticulum forces the food into the oesophagus where a rapid antiperistaltic wave propels it upwards into the mouth where the food is moistened with saliva, chewed and swallowed again. The time spent ruminating depends on the quality of the food. A sheep fed on fine hay will ruminate five hours out of 24 and one fed on coarse hay will ruminate for nine hours. Eructation of ruminal gases involve only movements of the rumen.

Control

Movements of the reticulo-rumen are controlled by the parasympathetic system. Sectioning the vagal nerves abolishes contraction; stimulation of these nerves causes contraction. Sectioning the sympathetic nerves has no effect on contraction; stimulation inhibits contraction. The intrinsic nerve supply is also important, for local reflex responses may be mediated through stretch receptor fibres in the walls. The strength and frequency of the contractions depend on the food content in the abomasum, and will decrease as food accumulates there.

Control of abomasal secretory activity is probably similar to the gastric phase of control in the monogastric stomach (p. 33). 'Sham feeding), however, does not stimulate abomasal secretion. The continuous secretion of abomasal juice depends mainly on the entrance of omasal contents into the abomasum and upon the VFA concentration. Copius secretion replaces fluid absorbed in the omasum. Secretion will decrease as the pH in the abomasum falls and as the acid chyme reaches the duodenum. Abomasal contractions are not co-ordinated with those of the other chambers of the stomach.

5   The pancreas

Structure

The pancreas is a large tapering gland lying in between the stomach and the duodenum. It is connected to the duodenum by the *pancreatic duct*. It is both exocrine and endocrine. The exocrine portion consists of *acini* joined by branching ducts that meet to form a main duct. The acini secrete several digestive enzymes (Chapter 7). The endocrine portion consists of small spherical granular cell groups (*islets of Langerhans*) which account for only one to two per cent of the total cell mass. These cells liberate two polypeptide hormones, *insulin* and *glucagon,* into the circulation, which respectively raise or lower blood glucose levels; both may affect alimentary tract secretions. Occasional goblet and enterochromaffin (or endocrine) cells are also seen in the islets.

Like the alimentary canal, the pancreas is innervated by the autonomic nervous system. Preganglionic vagal fibres synapse in ganglia within the gland, from which postganglionic fibres innervate the acinar and islet cells and smooth muscles of the ducts. Sympathetic fibres synapse within the coeliac ganglion, from which fibres innervate the pancreatic blood vessels.

The pancreas is richly supplied with blood capillaries. Oxygenated blood comes from branches of the coeliac and superior mesenteric arteries which form a capillary bed around the acini and islets. Blood is drained from the pancreas by two branches of the portal vein to the liver.

Secretions

Pancreatic juice is a colourless, isotonic, bicarbonate-rich solution containing many of the extracellular enzymes of the alimentary tract. In man about one and a half litres of alkaline fluid (pH 8.0–8.5) are emptied daily into the duodenum, which neutralizes the acid food entering the duodenum from the stomach and allows the enzymes in the pancreatic juice to act optimally in a neutral or slightly alkaline medium.

Pancreatic enzyme precursors are synthesized in the pancreas by acinar cells on ribosomes attached to the rough endoplasmic reticulum. They are then transported across the membrane of the reticulum into reticular cavities (cisternae) where they are concentrated. The precursor enzymes then move into small vesicles near the Golgi apparatus where they are bound by smooth-surface membrane and become granular (zymogen granules). The vesicles probably coalesce as they move towards the apical surface of the acinar cells, a process probably

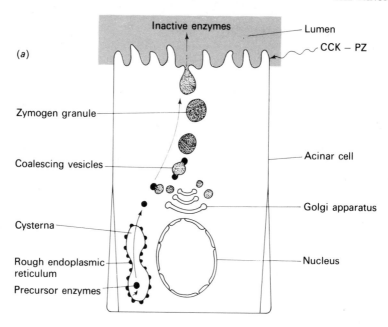

(a)

Inactive enzymes

Lumen

CCK – PZ

Zymogen granule

Coalescing vesicles

Acinar cell

Golgi apparatus

Cysterna

Nucleus

Rough endoplasmic reticulum

Precursor enzymes

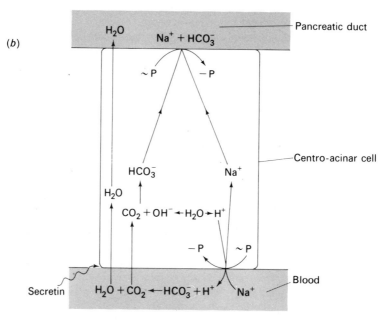

(b)

Pancreatic duct

H_2O

$Na^+ + HCO_3^-$

$\sim P$

$-P$

Centro-acinar cell

HCO_3^-

Na^+

H_2O

$CO_2 + OH^- \leftarrow H_2O \rightarrow H^+$

$-P$

$\sim P$

Blood

Secretin

$H_2O + CO_2 \leftarrow HCO_3^- + H^+$

Na^+

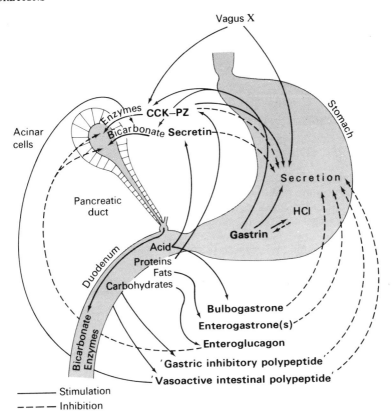

Fig. 5.2   Control of pancreatic secretion. Effects of hormones on gastric section, but not motility, also shown (depending on the species CCK-PZ will either decrease or increase gastric secretion).

Fig. 5.1   (a) Intracellular pathway of pancreatic enzyme precursors. CCK-PZ causes a release of zymogen granules from the acinar cells by an unknown mechanism (modified from Texter et al., 1968). (b) Possible mode of secretion of bicarbonate from the cells lining the pancreatic intralobular ducts. H$^+$ ions are actively removed from the base of the centro-acinar cell in exchange for Na$^+$ ions in the plasma. Water and CO_2, derived from plasma HCO_3^- ions, diffuse into the cell. The enzyme carbonic anhydrase in the cell hydrates CO_2 to form HCO_3^-. NA$^+$ and HCO_3^- ions are actively extruded from the apical surface. Secretin is thought to increase the permeability at the base of the cell (adapted from Scratcherd, T. and Case, R. M. (1973) Am. J. clin. Nutr. 26: 326–39).

requiring energy, and fuse with the cell membrane. Following stimulation of the cells by the hormone CCK–PZ (below) and the vagal nerve, the accumulated zymogen granules become discharged by an unknown but energy-requiring mechanism (Fig. 5.1a). The action of CCK-PZ is enhanced by the added presence of secretin.

Pancreatic enzymes taking part in digestion are secreted in an inactive form, to be activated by other enzymes and ions in the duodenum; their functions will be considered later (Chapter 7).

Control

Pancreatic secretions are mainly controlled by hormones produced in the anterior region of the small intestinal mucosa in response to the composition of the duodenal food, and to a smaller extent by nerves (Fig. 5.2). Acid chyme from the stomach, and to a lesser extent amino acids and fatty acids, stimulate the release of *secretin* from small endocrine cells (S-cells) found mainly in the transitional zone between the villi and crypt cells of the duodenum and upper jejunum (the stomach may also synthesize this hormone). Secretin was the first alimentary tract hormone to be discovered in 1902 by W. M. Bayliss and E. H. Starling. This hormone reaches the pancreas by the circulation and elicits the release from the pancreatic (intralobular) ducts of a watery solution rich in bicarbonate and other electrolytes but poor in enzymes (Fig. 5.1b). Secretin also stimulates the flow but not the control of bile from the liver cells, and slows gastric emptying and secretion. 'Vasoactive intestinal polypeptide' synthesized in the upper small intestinal mucosa will also stimulate the pancreas to secrete pancreatic juice low in enzyme-content, although its activity is only about one-tenth that of secretin.

Originally, it was thought that gall bladder contraction was caused by the release of *cholecystokinin* and pancreatic secretion caused by the release of *pancreozymin,* but now both hormones are regarded as one and called *cholecystokinin-pancreozymin,* or *CCK–PZ.* This hormone is released from the duodenal and proximal jejunal mucosa (endocrine I or X-cells) under the stimulus of partially digested proteins. (CCK-PZ has also been isolated in the pyloric antrum.) Certain amino acids, emulsified fatty acids and hydrochloric acid are less effective; vagal stimulation and bile salts may also affect its release. CCI-PZ stimulates the secretion of a moderate volume of pancreatic juice rich in enzymes from the acinar cells and, like gastrin, increases gastric acid secretion. Both CCK-PZ and secretin also cause vasodilation of the intestinal blood vessels. Carbohydrates in a meal have no effect on the release of either secretin or CCK-PZ.

There is also a gastric phase of pancreatic secretion. Distension of the pyloric antrum by the food and the presence of proteins will reflexly release *gastrin,* which will cause a scanty flow of pancreatic juice containing some enzymes and bicarbonate.

Pancreatic juice secretion is partly under the control of the autonomic nervous system. The sight, smell or taste of food may induce an initial slight flow of pancreatic secretion rich in enzyme-content. Secretion is mediated mainly by the *vagal nerve*. 'Sham feeding' a dog has a similar effect.

Little is known about the factors that inhibit pancreatic secretion. *Enteroglucagon,* a hormone released from the fundus of the stomach (A cells) and jejunal mucosa (L cells) into the bloodstream in response to mainly carbohydrates, reaches the exocrine pancreatic cells and inhibits pancreatic secretion. 'Gastric inhibitory polypeptide' from the pyloric and small intestinal mucosa (D, cells) may have a slight inhibitory effect on pancreatic fluid secretion but not on pancreatic enzyme output (enteroglucagon and 'gastric inhibitory polypeptide' are possibly identical). Hypertonic solutions in the lumen of the anterior alimentary canal will also depress enzyme output from the pancreas by an unknown mechanism.

6 Bile

Composition

Bile is a complex, neutral, isotonic solution, containing 97 per cent water and three per cent solids. In man, about half a litre of bile is secreted every 24 hours. The solids include bile pigments, bile salts, bile acids and electrolytes, mainly sodium chloride and bicarbonate. The main bile pigment is bilirubin, a breakdown product of haemoglobin in old red blood corpuscles. Bilirubin is converted by bacteria in the large intestine into urobilinogen, which becomes oxidized to urobilin, and excreted in the faeces. Most of the bile salts, mainly cholic, deoxycholic and taurocholic acids, are absorbed by the distal small intestine and re-secreted into the liver.

Bile is synthesized by the liver cells and secreted into small ducts uniting to form a common duct, the *bile duct*. The bile duct drains into the duodenum, although it may connect with the pancreatic duct shortly before doing so. Bile from the liver is stored in a small pear-shaped sac, the *gall bladder,* found on the underside of the liver. The gall bladder is missing in some animals as, for example, the rat. In the gall bladder the concentration of bile is increased about tenfold by the active re-absorption of salts and water across the gall bladder epithelium. Bile contains neither enzymes nor hormones, but bile salts are important in the digestion and absorption of fats in the food.

The gall bladder is innervated by a branch of the vagus (parasympathetic) and by fibres from the coeliac ganglion (sympathetic). Oxygenated blood is supplied by a branch of the coeliac artery and deoxygenated blood removed by a branch of the portal vein.

Control

Bile secretion is controlled by humoral, neural and, especially, chemical pathways (Fig. 6.1). Bile is prevented from escaping from the gall bladder by a sphincter, the sphincter of Oddi, found at the base of the bile duct where it enters the duodenal wall. Bile is stored and becomes concentrated in the gall bladder in between meals.

Shortly after food enters the stomach the gall bladder contracts, the sphincter relaxes, and bile flows into the duodenum. The hormone *CCK–PZ* is responsible for eliciting the contraction of the gall bladder wall, increasing bile secretion, and relaxing the sphincter. Secretin, gastrin and 'vasoactive intestinal polypeptide' increase the volume of bile produced by the liver but not the output of bile salts.

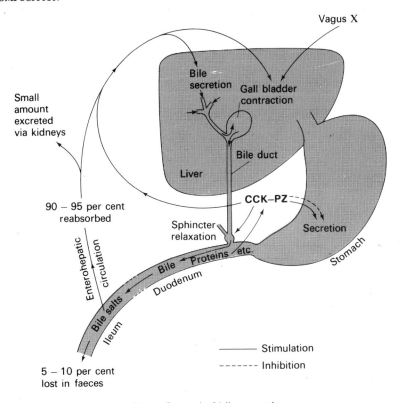

Fig. 6.1 Control of bile secretion.

In addition, eating will initiate reflexes mediated by the *vagal nerve* supplying the liver which will contract the gall bladder and result in increased bile flow.

Finally, increased *bile salts* in the circulation will stimulate bile secretion. Bile salts themselves are the most important agents for increasing bile secretion. In the ileum bile salts are mostly re-absorbed and returned to the liver by the hepatic portal vein (enterohepatic circulation), thereby increasing the rate of bile flow. Bile salts re-circulate about eight times every 24 hours.

7   The small intestine

Structure

Digestion and absorption of food are completed in the small intestine (small bowel). Most of the fluid in the alimentary canal is also absorbed there. It is often stated that the small intestine secretes digestive enzymes but recent investigations have shown that this is incorrect. The low enzyme activity found in small intestinal juice originates mainly from damaged epithelial cells that are continually being extruded into the lumen. Intestinal glands (Brunner's glands and crypt cells) secrete mainly mucus, electrolytes and possibly amylase. Digestion in the small intestinal lumen is accomplished mainly by enzymes in the secretions of the pancreas (Table 7.1). The final stages of digestion are often intracellular, not extracellular, as is commonly supposed, and are fully discussed in later chapters.

The mucosa of the proximal small intestine is an important endocrine region and synthesizes at least 12 polypeptide hormones of relatively low molecular weight, unlike that of the stomach which produces mainly one.* The functions of most of the intestinal hormones are well-defined, several having overlapping activities (Table 7.2).

The small intestine is an extensive, hollow, coiled, muscular tube lying in between the pyloric and ileocaecal sphincters. In man, it is 400–600 cm long and about four cm wide. The small intestine is arbitarily divided into the duodenum, jejunum and ileum in the approximate ratio 1:4:7 respectively. Externally, the different regions look alike although proximally the intestine is slightly thicker and more vascular. The surface area of the duodenum and jejunum is about four times greater than that of the ileum. The mucosa consists of a continuous sheet of epithelium thrown into numerous folds or ridges proximally and leaf or finger-shaped distally, the *villi*, projecting into the intestinal lumen (Fig. 7.1). Humans living in the tropics, however, tend to have uniformly convoluted or leaf-shaped villi. In man, each villus is about 1.0 mm tall, varying in number from 20 to 40 per mm^2 in the ileum or duodenum respectively. The villi are capable of movements that are controlled by a hormone, *villikinin*, released from the duodenal mucosa into the circulation after the arrival of acid food in the lumen. Human gastrin also causes movements of the villi; most other gastrointestinal hormones inhibit their movements.

---

* Endocrine polypeptide-producing cells are sometimes referred to as APUD cells, that is, they are Aminergic, displaying amine Precursor Uptake and Decarboxylation properties.

Table 7.1 Action of the enzymes of the mammalian alimentary tract

Site of secretion	Enzyme	Site of action	Substrate acted upon	Products of action
Mouth	Salivary α-amylase	Mouth	Starch	Disaccharides (few)
Stomach	Pepsin	Stomach	Proteins	Large peptides
Pancreas	Pancreatic α-amylase	Small intestine	Starch	Disaccharides
Pancreas	Trypsin	Small intestine	Proteins	Large peptides
Pancreas	Chymotrypsin	Small intestine	Proteins	Large peptides
Pancreas	Elastase	Small intestine	Elastin	Large peptides
Pancreas	Carboxypeptidases	Small intestine	Large peptide	Small peptides (oligopeptides)
Pancreas	Aminopeptidases	Small intestine	Large peptides	Oligopeptides
Pancreas	Lipase	Small intestine	Triglycerides	Monoglycerides, fatty acids, glycerol
Pancreas	Nucleases	Small intestine	Nucleic acids	Nucleotides
(Small intestine	Enterokinase	Small intestine*	Trypsinogen	Trypsin)
Small intestine	Disaccharidases	Small intestine*	Disaccharides	Monosaccharides
Small intestine	Peptidases	Small intestine*	Oligopeptides	Amino acids
Small intestine	Nucleotidases	Small intestine*	Nucleotides	Nucleosidases, phosphoric acid
Small intestine	Nucleosidases	Small intestine*	Nucleosides	Sugars, purines, pyrimidines

* Intracellular

Table 7.2 Action of hormones of the mammalian alimentary tract*. The existence of the last two hormones is uncertain

Source	Hormone	Stimulus to secretion	Action
Pyloric antral mucosa; duodenal mucosa	Gastrin	Protein and oligopeptides in stomach; antral distension; vagus	Stimulates gastric acid secretion. Stimulates some enzyme and weak bicarbonate output by the pancreas. Stimulates bile flow.
	Secretin	Acid (mainly), oligopeptides and fats in duodenum and proximal jejunum. Local reflexes and vagus	Stimulates secretion of watery alkaline, enzyme-poor pancreatic juice. Inhibits gastric secretion and motility. Increases small intestinal secretion. Stimulates bile flow.
	Cholecystokinin—pancreozymin (CCK–PZ)	Oligopeptides (mainly) and fats in duodenum and proximal jejunum. Vagus and bile salts	Contracts gall bladder and stimulates bile secretion. Stimulates secretion of enzymes from pancreatic (acinar) cells. Inhibits/stimulates gastric secretion and motility.
	Enterogastrone(s)	Fats (mainly) in jejunum	Inhibits gastric secretion and motility.
	'Gastric inhibitory polypeptide'	Monosaccharides and fats in duodenum	Inhibits gastric secretion and motility. Stimulates secretion of Brunner's glands.
Upper small intestinal mucosa	Bulbogastrone	Acid food in duodenum	Inhibits gastric secretion and motility.
	Enteroglucagon	Carbohydrates (mainly) in duodenum	Inhibits jejunal motility and pancreatic secretion.
	'Vasoactive intestinal polypeptide'	Fats in duodenum (?)	Increases blood flow. Stimulates small secretion of pancreatic fluid, low in enzyme content. Inhibits gastric secretion.
	Motilin	Alkaline food in duodenum	Stimulates gastric and small intestinal motility.
	Villikinin	Acid food in duodenum	Stimulates movements of the villi; no action on muscular coat.
	Duocrinin	Acid food in duodenum	Enhances secretion of Brunner's glands.
	Enterocrinin	Acid food in duodenum	Stimulates secretion of the crypts of Lieberkühn.

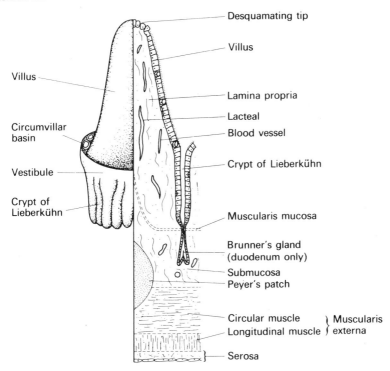

Fig. 7.1   Generalized structure of the small intestine, showing details of a villus and a crypt.

Around the base of each villus, and with epithelium continuous with that of the villus, are several simple, narrow (diameter 100 μm) indentations or pits, the *crypts of Lieberkühn*. The crypts open into a *vestibule*, common to four or five crypts. Four to ten vestibules coalesce to form a *circumvillar basin* around the base of each villus. The proximal intestine contains about twice as many crypts as the distal intestine. Although the number of villi remain constant throughout life, the number of crypts increase two-fold by longitudinal fission. The ratio of villus to crypts varies in different animals. For example, the ratio averages 1 to 5.5 in the mouse and 1 to 13 in the rat.

The epithelium of the villus consists of a single layer of cylindrically-shaped *absorptive cells*, whose surface area is enormously increased by minute processes, the *microvilli*. Occasional *goblet cells* are interspersed among the absorptive cells. They secrete mucus that lines the epithelial surface, preventing both mechanical damage by the food and invasion by bacteria in the lumen. Rarely, *enterochromaffin cells* are also identified. Three different types of enterochromaffin-like *endocrine cells* have been isolated, mainly in the proximal small intestine; these have been little studied. Small lymphocytes *(thelio-lymphocytes)* are sometimes found packed within the intercellular epithelial spaces; these cells have no known functions. Lining the base of the crypts are *Paneth cells* and along the sides *undifferentiated cells* and enterochromaffin cells (their detailed structure will be given in Chapter 9). The epithelial cells have a high metabolic rate concerned with active intracellular events. The epithelium rests on a thin *basement membrane.*

The central core of the villus, or lamina propria, consists of loose connective tissue, strands of smooth muscle passing upwards from the muscularis mucosa, nerves, leucocytes, an arteriole and venules forming a rich capillary network, and a central lymphatic vessel, the lacteal. Beneath the muscularis mucosa and submucosa are the typical muscular and serous coats. Lymphoid nodules, or *Peyer's patches,* extend in the mucosa and submucosa at intervals along the small intestinal length. The small intestine is attached to the body wall by a thin sheet of tissue, the *mesentery.*

Parasympathetic innervation is by a branch of the vagus and sympathetic innervation by sympathetic nerves from the superior mesenteric ganglion. Oxygenated blood is supplied via the abdominal aorta by branches of the coeliac and superior mesenteric arteries, and deoxygenated blood removed to the portal vein by the splenic and superior mesenteric veins. Over half the total small intestinal blood flow is directed to the mucosa and up to one fifth distributed to the villi.

Secretions

The intestinal juice *(succus entericus)* is a viscous mucus secreted by Brunner's glands and by cells lining the crypts of Lieberkühn. About one and a half litres of fluid are secreted by man every 24 hours. Originally it was thought that Brunner's glands, which are ramifying tubules lying in the submucosa of the duodenum, secreted several enzymes through ducts opening into the crypts, but recent evidence indicates that the glands secrete a viscid, slightly alkaline (pH 7.0–8.0), largely enzyme-free fluid, containing mainly mucoproteins and bicarbonate.

The secretions of Brunner's glands are believed to be controlled by several hormones. Acid food reaching the duodenum provokes endocrine cells in the duodenum to release *secretin* and *duocrinin* that reach Brunner's glands by the circulation. These hormones, together with CCK-PZ, gastrin, 'gastric inhibitory polypeptide' and 'vasoactive intestinal polypeptide' possibly enhance secretion. However, their effects on the secretion of Brunner's glands are conflicting. The functions of the alkaline fluid are unclear, although it undoubtedly protects the lining of the proximal small intestine from acid food passed on from the stomach and serves to lubricate the inner surface.

Undifferentiated cells in the crypts of Lieberkühn secrete a clear, alkaline intestinal fluid into the lumen. Their secretion is probably under both nervous and humoral control. Distension of the duodenal wall elicits local reflex stimuli that stimulate the secretory cells. Vagal stimulation also causes increased secretion. Acid food entering the duodenum stimulates the duodenal mucosa to produce *enterocrinin*, which possibly causes the crypt cells to secrete fluid.

The mode of action of some of the small intestinal hormones and the nature of the secretions of the small intestine are far from clear. Current studies indicate that other hormones synthesized by the small intestinal mucosa are likely to be discovered soon*. Moreover, only five out of 11 endocrine cell-types have been clearly shown to synthesize different hormones.

Digestion

Although the small intestine does not secrete digestive enzymes, there is considerable digestive activity in its lumen or on its surface. These enzymes are secreted by the pancreas, delivered into the duodenum through the pancreatic duct, and have proteolytic, lipolytic and amylolytic activities (Table 7.1). Their actions in monogastrics are given below.

Proteolytic activity

Gastric pepsin is relatively inefficient and soon rendered inactive when the food enters the duodenum. Over three-quarters of the total enzyme output in the pancreatic juice acts on proteins. The enzymes *(proteases)* are synthesized in the pancreatic acinar cells and secreted as inactive precursors, or *proenzymes,* into the duodenal lumen where they become activated (p. 47).

(a) The inactive precursor *trypsinogen* is converted to *trypsin* by an enzyme, *enterokinase,* located on or within the apical surface of the duodenal villi. (Enterokinase is probably released under the stimulus of bile acids and cholecystokinin pancreozymin.) (b) *Chymotrypsinogen(s)* is also inactive and is activated by trypsin to *chymotrypsin(s).* (c) *Proelastase* becomes activated by trypsin to *elastase* which specifically attacks the fibrous protein elastin. These enzymes are all *endopeptidases (proteinases),* splitting inner peptide (CO–NH) bonds between adjacent amino acids into large peptide molecules. (d) *Procarboxypeptidase A* and *B* are inactive and become converted by pepsin or enterokinase to *carboxypeptidase A* and *B.* These are *exopeptidases (peptidases)* that split off terminal peptide bonds adjacent to carboxyl (—COOH) groups, liberating small peptide molecules and a few amino acid molecules. (e) *Aminopeptidases,* a multiple group of exopeptidase enzymes, will cleave terminal peptide links adjacent to α-amino (—NH$_2$) groups, also releasing small peptide molecules. Recent evidence suggests that aminopeptidases are probably mainly located within the small intestinal epithelium.

Many small, diverse peptide molecules *(oligopeptides)* are not digested in the lumen of the small intestine. Their eventual fate, and that of free amino acids in the food, will be discussed later (Chapter 13).

* See, for example, Grossman, M. I. *et al.* (1974). Candidate hormones of the gut. *Gastroenterology* **67,** 730–55.

Pancreatic *nucleases* (four per cent of total pancreatic enzymes) split nucleic acids into nucleotides, which are split into nucleosides and phosphoric acids by *nucleotidases* in the intestinal juice. Nucleosides are converted into purine and pyrimidine bases and sugars by intestinal *nucleosidases*.

Lipolytic activity

Compared with the proteolytic enzymes, lipolytic activity in pancreatic juice is low. Pancreatic *lipase (steapsin)* hydrolyzes fats, which are emulsified by bile salts in the duodenum. Although lipase has also been isolated in the saliva and gastric juice, virtually all dietary fats are digested and absorbed in the small intestine. Triglycerides are hydrolyzed to diglycerides, and diglycerides hydrolyzed to monoglycerides, free fatty acids and glycerol. Their subsequent fate will be discussed later (Chapter 14).

Amylolytic activity

Pancreatic α-*amylase (amylopsin)* is similar to salivary α-amylase and will also split polysaccharides (starch) into a mixture of different disaccharides. Since the amylolytic activity in pancreatic juice is low, presumably amylases within the small intestine epithelium are important in hydrolyzing polysaccharides. The final stages of digestion, when specific disaccharidases hydrolyze different disaccharides into monosaccharides, will be discussed later (Chapter 12).

Movements

The activity of the small intestine has been compared to that of a cocktail shaker. As the food reaches the duodenum, the small intestinal wall begins a complicated series of contractions, relaxations, to-and-fro movements and peristaltic waves that knead the intestinal contents and push the food distally. The resulting creamy consistency is known as *chyle*. Different species of mammals show different kinds of intestinal movements; it is rare for any one species to show all types.

Intestinal contractions are weak or missing during fasting but increase following a meal. Periods of intestinal motility alternate irregularly with periods of inactivity. Small pressure waves of irregular duration and amplitude *(phasic waves)* occur infrequently and are usually grouped together (Fig. 7.2). The purpose of such waves is to mix the food with the digestive juices and to bring it into contact with the absorptive epithelium. These waves are more frequent in the upper intestine than in the lower intestine.

Mixing of intestinal contents is also accomplished by the formation of characteristic small, ring-like contractions to form 'segmentation' of portions of the intestine. Segmentation can be haphazard or orderly. Contraction waves (pendular movements) may also pass back and forth for short distances along the intestine to help the mixing process.

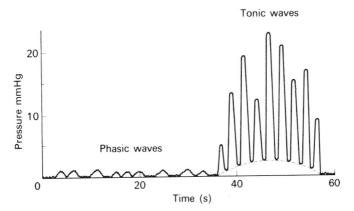

Fig. 7.2   Phasic (Type I) and tonic waves (Type III) recorded from the mammalian small intestine. Phasic waves each last between 2–12 s, with an amplitude of under 5 mmHg, and occur irregularly. Tonic waves, with an elevated base-line, have an amplitude of 5–20 mmHg, and last from 10 s to several minutes (modified) from Texter *et al.*, 1968).

Such mechanisms, however, provide little net propulsion of the chyle. The food is pushed distally mainly by peristaltic waves *(tonic waves)* of greater amplitude and duration (Fig. 7.2). They may start in any region of the small intestine and travel for only short distances. The rate of movement of the food along the intestine resulting from such contractions is slow, allowing the food time to be digested and absorbed.

Undigested and unabsorbed food reaches the ileocaecal sphincter at the distal end of the intestine about three hours after entering the duodenum. This sphincter is normally closed. As the food enters the stomach, peristaltic waves in the distal intestine increase and relax the ileocaecal sphincter, allowing chyle into the caecum and large intestine. This mechanism is triggered by a gastro-ileal reflex action. The sphincter then contracts to prevent regurgitation of the large intestinal contents into the small intestine.

Control

Small intestinal movements are determined mainly by various local neuromuscular reflex mechanisms, but larger movements are regulated through extrinsic nerves. The main intrinsic innervation is the submucous and myenteric nerve plexuses, originating from extrinsic nerves consisting of a branch of the vagus (parasympathetic) and sympathetic nerves from the superior mesenteric ganglion. Mild distension of the inner surface of the small intestine by the food will stimulate mechano- and chemoreceptors and increase intestinal motility by local peristaltic reflex mechanisms.

Intestinal slow (phasic) waves are generated from longitudinal muscles to circular muscles via muscle junctions (p. 13). Such responses are independent of the extrinsic innervation, for if the vagal or sympathetic nerves are cut intestinal motility remains unaffected. Peristaltic (tone) waves, however, are co-ordinated by the central nervous system via mainly afferent vagal impulses. Possibly the hormones CCK–PZ, gastrin and motilin also increase small intestinal motility, while enteroglucagon and secretin possibly inhibit motility.

8 The large intestine and rectum

Structure

The *large intestine* (colon, large bowel) is less than half as long but about twice as large as the small intestine. It consists of three straight segments forming ascending, transverse and descending portions. The descending portion bends distally (the *sigmoid flexure*) and ends as a short, straight tube, the *rectum.* The external opening or *anus* is closed by two *anal sphincters,* the *internal* and *external.* In carnivores, the *caecum* is a blind-ended pouch lying below the junction of the small and large intestines and, except for some absorption of water and electrolytes, is non-functional. However, in non-ruminant herbivores (for example, the horse, pig, rabbit and guinea-pig) both the caecum and large intestine are greatly enlarged. This is to provide a milieu suitable for further digestion of food, which is carried out by bacteria and protozoa in the lumen and by the action of pancreatic enzymes mixed with the food. The *appendix,* a finger-like projection from the caecum has, in man, no known function. Following a meal, the *ileocaecal sphincter* opens at frequent intervals and the contents of the small intestine and caecum pass into the large intestine.

The large intestine lacks digestive enzymes. However, the large intestine of the dog and mouse, but not that of man, rat and guinea-pig, is able to actively absorb sugars and amino acids, like the small intestine. The proximal portion absorbs sodium and chloride and the products synthesized by bacteria, and secretes potassium and bicarbonate. In man, it receives daily up to one litre of fluid, much of which is slowly absorbed, producing a rise in dry matter, the *faeces.* The distal portion, which stores and controls the expulsion of faeces, contains undigested food residues and a large population of gas-producing bacteria.

The mucosa has a wavy surface lined by a layer of simple columnar epithelial cells, occasional goblet cells and a few enterochromaffin cells (Fig. 8.1). Numerous epithelial tubules *(glands of Lieberkühn),* arranged perpendicularly to the outer surface, dip into the mucosa. Each tubule is lined by goblet cells and the base contains undifferentiated cells and enterochromaffin cells. The goblet cells cover the epithelial surface with a thin layer of protective mucus. The undifferentiated cells divide and proliferate from the base of the glands and migrate to the surface in two to eight days, differentiating as they do so. *Lymph nodes* may occasionally be identified in the submucosa. Both the muscularis externa and serosa are typical, although the outer longitudinal muscle layer is sometimes, as in man, concentrated into distinct bands (the *taeniae coli*), folding the walls into sacs (*haustra*) by the contraction of a continuous inner layer of circular muscle.

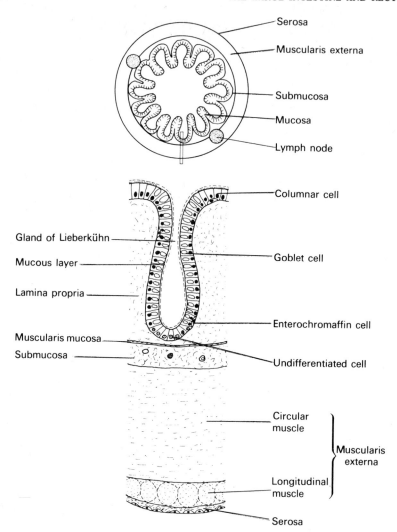

Fig. 8.1   Transverse section through the large intestine of a mammal (e.g. rat).

The ascending and transverse limbs of the large intestine are innervated by branches of the parasympathetic vagal nerve, while the descending limb, rectum and internal anal sphincter are supplied by the parasympathetic pelvic (sacral) nerve. The sympathetic system consists of several branches, one from the superior mesenteric ganglion to the ascending portion, and others from the inferior mesenteric ganglion to the transverse and descending portions, rectum and internal anal sphincter. In addition, the transverse and descending portions and the external anal sphincter are innervated by part of the somatic nervous system, the pudential nerve. Oxygenated blood from the abdominal aorta is supplied by the superior and inferior mesenteric arteries and deoxygenated blood removed to the liver by the portal vein via the superior and inferior mesenteric veins.

Secretions

About half a litre of secretions from the large intestine is produced in man every 24 hours. The secretions consist of a scanty, enzyme-free, watery alkaline (pH 7.5–8) fluid with clumps of mucus, and bicarbonate and potassium ions. Most of the remaining luminal water and sodium chloride are absorbed. The mucus binds the faecal matter and lubricates the inner surface. Local mechanical stimulation, initiated by the parasympathetic pelvic nerves, increases secretion; stimulation of the sympathetic nerves reduces secretion.

Movements

Contractions of the distal small intestine push undigested food into the caecum which fills passively, and the food is gradually propelled by caecal contractions into the ascending limb of the large intestine through the ileocaecal sphincter. Food entering the stomach or duodenum will also cause marked contractions in the ascending limb. Food residues little by little accumulate in the ascending limb, and slow segmental contractions eventually push the food mass into the transverse limb. To-and-fro movements (tone waves) caused by contractions of the longitudinal muscles, are also evident. Materials are suddenly propelled down the empty descending limb, one to three times daily, by strong contractions of the ascending and transverse portions. These contractions are caused by a gastro-colonic reflex action, usually after a meal (or, apparently, by listening to a lecture on defaecation), which eject the faeces into the rectum immediately before evacuation. Slow contraction waves of low amplitude (phasic waves), which are generated in the longitudinal muscle layer, have been recorded in the distal large intestine and rectum; the physiological significance of such waves is uncertain as they are probably not propulsive. Although the hormones CCK–PZ, gastrin and secretin can affect movements of the large intestine, it is not known whether they serve in regulation. The gastro-colonic reflex is possibly controlled by gastrin.

Defaecation

The animal now has the privilege of controlling its alimentary canal function which left it when the food reached the pharynx (Fig. 8.2). The rectum is normally closed by two anal sphincters, the internal sphincter which is composed of circular smooth muscle and innervated by the autonomic nervous system, and the

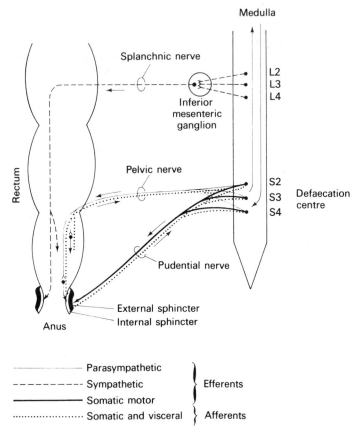

Fig. 8.2   Nervous regulation of defaecation (details in text). (Modified from Monnier, M. (1968) *Functions of the Nervous System*, p. 404). Publisher: Elsevier Pub. Co. Ltd. Amsterdam.

external sphincter which is composed of striated muscle and innervated by the voluntary (somatic) nervous system.

Marked distension of the walls of the rectum produced by sudden movement of faecal matter in the lumen stimulates stretch receptors there. Afferent impulses in the pelvic nerve (parasympathetic) activate the defaecation centre in the sacral cord. Efferent impulses in the rectal wall inhibit tonic contractions and relax the internal sphincter, inducing a reflex opening.

The voluntary control of defaecation means that defaecation may be helped or stopped at will. The act involves the external sphincter, and is co-ordinated and controlled by the defaecation centre in the medulla via the pudential nerve. Impulses from the external sphincter are conducted to the medulla by somatic afferent fibres and returned to the external sphincter by somatic efferent fibres, which elicit the opening of the external sphincter. Evacuation is assisted by inspiration, followed by contraction of the chest and abdominal muscles.

Defaecation is inhibited by efferent impulses from the sympathetic innervation that originate from the lumbar cord and end in the rectal muscles and internal sphincter.

Faeces

Faeces consist of unabsorbed food, dead bacteria and fluid. The solid fraction includes bacteria (30 per cent), cellulose and roughage (30 per cent), fat (10–20 per cent), protein (2–3 per cent), mucosal cells and mucus. The inorganic fraction (15–20 per cent) contains mainly calcium, magnesium and potassium, with smaller amounts of iron, sodium, bicarbonate and sulphate. The faeces are normally about three-fourths fluid and one-fourth solid.

II   Absorption

Where sciences meet, their growth occurs. In scientific borderlands not only are facts gathered that are often new in kind, but it is in these regions that wholly new concepts arise. (F. Gowland Hopkins)

THE SUPERBOWEL*

I think that I shall never see
A tract more alimentary.
A tube whose velvet villi sway
Absorbing food along the way.
Whose surface folded and striate
Does rapidly regenerate.

A magic carpet whose fuzzy nap
Miniscule molecules entrap.
Then, microvilli with enzymes replete
The last hydrolyses complete.

A tunnel studded with protection
Against abrasion and infection
(Goblets their mucus spill
While lymphoid cells the microbes kill).

To top things off, it should be noted,
This Grand Canal is sugar coated!

(George J. Fruhman)

* Reproduced by permission from *Perspect. Biol. Med.* (1973) **17**, 66.

9 Villous cell structure

The structure of the small intestine as seen under the light microscope (Fig. 7.1) has already been described. In this chapter the detailed structure of the main types of cells of the villus and crypts, particularly that of the absorptive cell, is given, as seen under the transmission electron microscope. The development of the small intestine and the renewal of the villus cells are also briefly outlined.

The absorptive cell *(enterocyte)*

Figure 9.1 illustrates the structure of the absorptive cell. It is columnar in shape and, in man, about 30 μm tall and 12 μm wide. In 1932 C. C. Macklin was the first to notice a fine striated or brush border covering the apical surface of the cell. In 1949 B. Granger and R. F. Baker showed that the striated border consisted of fine projections, the *microvilli*. There are 2000 to 3000 microvilli, each 1 μm long and 0.1 μm wide, covering the entire outer surface of each cell, with intermicrovillous spaces of 0.05 μm. Along the base of the crypt, however, the microvilli are shorter and wider (0.2 μm × 0.2 μm) but gradually increase in length and become narrower from the bottom of the crypt upwards.

The microvilli enormously increase the surface area of the intestine. In man, the cylindrical small intestinal surface area is about 0.4 m². Surface projections *(folds of Kerckring)*, on which the villi rest, increase the surface area to 1 m². The surface area of the villi is 10 m², and that of the microvilli at least 200 m².

The outer surface of the microvilli is covered by a fine mat of branching filaments, up to 0.3 μm thick, the *glycocalyx (fuzz or enteric surface)*, first observed in 1963 by H. S. Bennett and described by S. Ito in 1964. The filaments are radially orientated, 25–50 Å in diameter, and their bases are continuous with the outer surface of the plasma membrane (Figs. 9.2 and 9.3). The glycocalyx is composed of acid mucopolysaccharide and glycoprotein and differs from the adherent mucus produced by neighbouring goblet cells. The glycocalyx is first formed from precursor monosaccharides in the cytoplasm which become fixed to peptide receptors on ribosomes attached to the rough endoplasmic reticulum. The material is then transferred to the Golgi apparatus where glycoproteins are synthesized, to be transported by the smooth endoplasmic reticulum to the plasma membrane of the microvilli where they are ejected, by an unknown mechanism, to form a surface coat. Kinetic studies have shown that the glycocalyx is replaced within four hours. The glycocalyx is therefore not a static component of the cell surface but is continually synthesized by the cell and forms an integral part of the dynamic membrane. The glycocalyx of the villus is complete but that of the crypts is incomplete.

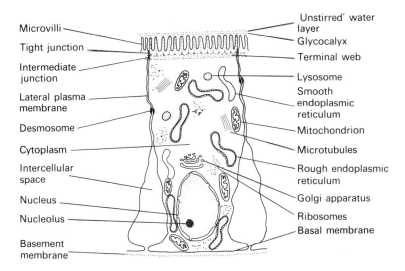

Microvilli

Tight junction

Intermediate
junction

Lateral plasma
membrane

Desmosome

Cytoplasm

Intercellular
space

Nucleus

Nucleolus

Basement
membrane

Unstirred' water
layer

Glycocalyx

Terminal web

Lysosome

Smooth
endoplasmic
reticulum

Mitochondrion

Microtubules

Rough endoplasmic
reticulum

Golgi apparatus

Ribosomes

Basal membrane

Fig. 9.1   The absorptive cell (adapted from Trier, 1968).

The glycocalyx may have several functions. It may act as a buffering layer over the microvilli against both physical and chemical agents, or present a selective barrier to the diffusion of certain molecules, or it may be the site of both the adsorption of extracellular enzymes and of surface (contact) digestion. The digestive functions of the glycocalyx will be discussed later (p. 106). Within the interstices of the glycocalyx and overlying the glycocalyx surface is a layer of *'unstirred' water* of varying thickness. There is obviously no sharp boundary between the water in the intestinal lumen and the 'unstirred' water surface, the static water layer blending imperceptibly with the water of the intestinal contents.

In 1967 C. F. Johnson isolated small particles or *knobs*, each 50 Å in diameter, with a space 20 Å in between each, attached by slender stalks to the outer surface of the microvilli. These knobs contain several intracellular enzymes; their functions will be given later (p. 107). It is possible, however, that these knobs are artefacts caused by clumping of filaments of the glycocalyx.

Each microvillous core contains a bundle of about forty protein *microfilaments* or, more probably, hollow *microtubules*, each 60 Å in diameter (Figs. 9.2 and 9.3). They run parallel to the long axis of the microvillus and end as rootlets in a feltwork of packed fibrils, about 0.4 μm wide, the *terminal web*, running parallel to the luminal surface of the cell. The function of the terminal web is to stiffen and stabilize the cell. Actin fibrils within the microvillous core suggest that the microvilli may have contractile properties.

The *plasma membrane* covering the cell surface is a trilamellar structure, con-

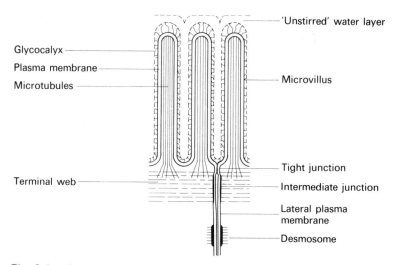

Fig. 9.2   Apical portion of the absorptive cell (modified from Trier, 1968).

sisting of electron dense outer and inner leaflets separated by a translucent middle leaflet (Fig. 9.3). The microvillous membrane is wider (100 Å) than either the basal or lateral membranes (85 Å each). *Pores* have never been seen piercing the microvilli. Recent studies have shown that the apical membrane of the absorptive cell is a site of intense biochemical activities, where the final stages of digestion and transport of materials take place. Both the fine structure and functions of the plasma membrane will be described in detail in the next chapter.

The *lateral plasma membrane* is infolded along the apical third of the cell and closely apposed to that of adjacent cells, while the basal two-thirds is only loosely adherent and sometimes well-separated to form prominent *intercellular spaces* (intercellular spaces are less obvious in the other cell-types). In between the base of the microvilli and the terminal web the plasma membrane forms an unthickened intercellular band, 0.1–0.2 µm long, encircling the cell apex, the *tight junction (zonula occludens);* it expands to 0.2–0.5 µm in length in the terminal web region, with a 200 Å gap in between, to form the *intermediate junction (zonula adhaerens).* Between the intermediate junction and the intercellular spaces the lateral plasma membrane forms several discrete disks, each 0.2–0.3 µm long, with a gap 240 Å in between, the *desmosomes (macula adhaerens)* that anchor adjacent cells by thin cytoplasmic filaments (Fig. 9.2).

Beneath the basal plasma membrane is the *basement membrane,* a deceptive term for it is not a membrane but an irregular flat layer of fine, homogeneous fibrils, 300 Å wide (the term *basal lamina* appears to be more appropriate). The basement membrance is possibly secreted in a similar manner to the glycocalyx. This layer is rich in carbohydrate-protein complexes synthesized by the cell. The basement membrane is in intimate contact with blood capillaries (endothelial cells), lymph channels and nerve fibres beneath it.

The remaining inclusions in the absorptive cell are like those of any other typical animal cell. The cell *cytoplasm* is a pale, amorphous material containing diverse elements, or *organelles,* and many enzymes. The *nucleus* is oval-shaped and located at the base of the cell. The *nuclear membrane* consists of two apposed membranes, each 75 Å thick, separated by a space. The membrane is pierced by numerous nuclear pores, each 500 Å in diameter and partly plugged by granular bodies, occupying 10–20 per cent of the total surface area of the nucleus. The nucleus contains most of the genetic material (DNA) of the cell and controls protein synthesis, cellular growth, and cellular reproduction (the mature absorptive cell, however, cannot divide). The nucleus contains one or more *nucleoli* composed of compact RNA granules. The nucleoplasm is diffuse; mitosis has not been observed in the mature absorptive cell. The *endoplasmic reticulum* is *rough* (that is, studded by ribosomes) and *smooth,* and both connect to form a netlike system of tubules distributed throughout the cell cytoplasm which are probably continuous with both the nuclear pores and the cell membrane. The rough endoplasmic reticulum is the site of many synthetic activities in the cytoplasm and facilitates intracellular transport of synthesized protein molecules from the ribosomes. The *Golgi apparatus (Golgi complex, Golgi body),* a stack of hollow tubules and budding vesciles, is inconspicuous and lies immediately above the nucleus. It is probably continuous with the endoplasmic reticulum and serves to synthesize, store, secrete, and transport metabolites (mucopolysaccharides and glycoproteins). The Golgi vesicles are thought to move towards and fuse with the plasma membrane, thus acting as a 'membrane factory'. Numerous *mitochondria* are scattered throughout the cell with their long axis parallel to that of the cell. They contain many oxidative enzymes that are involved in energy-yielding reactions in the cell. *Lysosomes* are confined to the apical half of the cell and contain acid hydrolytic enzymes and acid phosphatase, enclosed in membrane vesicles that prevent the cell from digesting itself. They provide a mechanism for digesting and eliminating obnoxious materials (bacteria, foreign proteins, ageing organelles) by lysis into innocuous materials. *Microtubules* are found throughout the supranuclear cytoplasm, aligned parallel to the long axis of the cell. They are possibly involved in intracellular transport and confer rigidity to the cell. Scattered free *ribosomes,* the site of protein synthesis, are located throughout the cell cytoplasm. Two *centrioles* are rarely seen in the apical cytoplasm.

A. M. Isomäki, (1973) has recently described a second type of absorptive-like cell, the *tuft cell,* scattered among the epithelial cells of the duodenal and gastric mucosa. Typically, tuft cells have a well-developed bundle of about ninety glycocalyx-covered microvilli projecting into the lumen. The plasma membrane surrounding the cell and intercellular junctions are like those of the absorptive cell. Microtubules in the microvillous core project deep into the apical cell cytoplasm; the terminal web is missing. The cell has a poorly-formed endoplasmic reticulum but a well-developed Golgi apparatus and many mitochon-

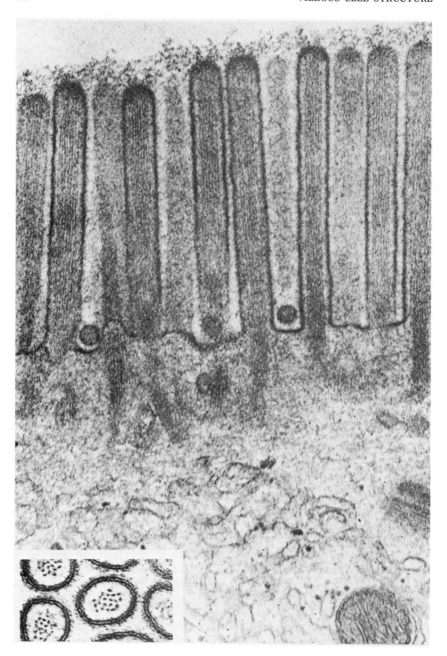

dria. Cytoplasmic membrane-cavities are obvious. There are many membrane-bound apical vesicles and cytoplasmic microtubules or microfilaments, glycogen granules and ribosomes. The nucleus is basal, and round or oval in shape. The functions of tuft cells are unknown, although their structure suggests that they are involved in selective absorption of macromolecules by pinocytosis, or they may function as detector cells.

The goblet cell

Goblet cells are occasionally found along the villi of the proximal small intestine but become more numerous distally. They account for 5–10 per cent of the epithelium. As the name suggests, the cell has the shape of a brandy goblet (Fig. 9.4a). The goblet cells proliferate from the crypts from undifferentiated and probably intermediate cells. The microvilli are short and irregular and the terminal web is poorly developed. Typically, the cell is distended by large, pale, mucous granules, each enclosed by a thin membranous coat. The mucous granules swell as water is imbibed by the cell, their membranous coat fuse together and with the apical plasma membrane, and the mucus is then extruded. Mucus is probably secreted continuously as it is synthesized. The mucus contains neutral, acidic and sulphur-containing glycoproteins and acid mucopolysaccharides. Prostaglandins (hydroxy fatty acids widely distributed in tissues) may have a stimulant action on the mucus-secreting cells. The mucus forms a lining over the surface of the villi (Fig. 9.5), distinct from the glycocalyx, and serves to lubricate the intestinal surface and deter bacteria from penetrating the cells. It also acts as a non-immunological anti-histamine defence mechanism. The nucleus is basal, and both the Golgi apparatus and endoplasmic reticulum are well-developed. Other organelles are similar to those of the absorptive cell.

The enterochromaffin cell

This cell is also called the *argentaffin cell*, since it is stained by silver salts, or the *intestinal endocrine cell*. Each is a small, triangularly shaped cell (Fig. 9.4b), abundant in the crypts and progressively decreasing in number towards the villous tip. They are numerous in the small intestinal epithelium of man, calf, and pig but are uncommon in rodents. Enterochromaffin-like cells are also found in the epithelium of other regions of the alimentary canal. Unlike other intestinal cells, the cell apex does not always reach the intestinal lumen and the cell thus

Fig. 9.3 Apical portion of the absorptive cell. The slender microvilli projecting into the lumen are coated with a glycocalyx. The surface of the microvilli consists of a trilamellar membrane (inset). Microtubules or microfilaments extend from the microvillous core into the terminal web. Transmission electron microscopy. Approximately ×57 740; inset approximately ×113 750. (Reproduced by permission from Rubin, W. (1971) *Amer. J. clin. Nutr.* **24,** 55).

sometimes lacks microvilli. The nucleus is irregularly-shaped and placed centrally. The cell cytoplasm has marked reducing properties, unlike the other intestinal cells. Numerous small membrane-bound, darkly staining, acidophilic secretory granules of different electron densities fill the cell, particularly near its base. They appear to be formed in the Golgi apparatus. The remaining organelles are also shared by the other cells.

The enterochromaffin cell is secretory, but its secretions are probably liberated into the sub-epithelial tissues rather than into the lumen of the crypts. The mechanism of secretion is unknown. The cell synthesizes a chemical substance, *serotonin* (5-hydroxytryptamine), which stimulates visceral muscle contraction and probably affects villous motility. Intracellular oligopeptides and lipoproteins have also been isolated.

Other enterochromaffin-like endocrine cells have been identified in the mucosa of the proximal small intestine in several species; they are thought to synthesize several hormones (Table 7.2). With the endocrine cells in the stomach, these cells sometimes have short microvilli projecting into the lumen which may act as detector devices. Yet other enterochromaffin-like cells have been found in the ileum; these cells have no clear function.

The undifferentiated cell

These cells are confined to the sides of the crypts where they are the most abundant of cell types (Fig. 9.4c). They proliferate rapidly by mitotic division into absorptive cells or goblet cells. The interphase cell is columnar in shape. When the cell is dividing, the nucleus is tri-lobed and usually placed centrally. The microvilli are sparse and irregular, and the terminal web underdeveloped. The lateral plasma membrane is straight, and the cell lacks intercellular spaces. The endoplasmic reticulum is poorly developed, and there are many free ribosomes. Small, membrane-bound secretory particles are found in the cell apex. Glycogen granules sometimes accumulate in the cytoplasm. Other organelles are like those of any typical cell.

Although the undifferentiated cell is secretory, the role of its secretions is unclear. As the undifferentiated cell migrates from the crypts onto the villus it loses its secretory activity and abruptly differentiates into an absorptive or a goblet cell. It is not known what controls cellular differentiation.

The Paneth cell

Paneth cells (Fig. 9.4d) line the base of the crypts where they persist for two to three weeks before disintegrating. Their frequency in different animals varies; they are abundant in man, ruminants and rodents but rare in domestic animals. They are more common distally than proximally. The cell is flask-shaped, with a broad base lying on a basement membrane. Microvilli are small and irregular, and the terminal web is ill-defined. The cytoplasm is basophilic and filled with large

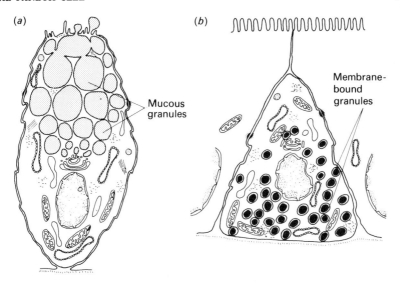

(a)

Mucous granules

(b)

Membrane-bound granules

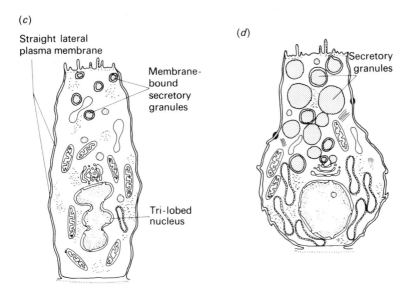

(c)

Straight lateral plasma membrane

Membrane-bound secretory granules

Tri-lobed nucleus

(d)

Secretory granules

Fig. 9.4   (a) The goblet cell. (b) The enterochromaffin cell. (c) The undifferentiated cell. (d) The Paneth cell. Details for a to d as in Fig. 9.1.

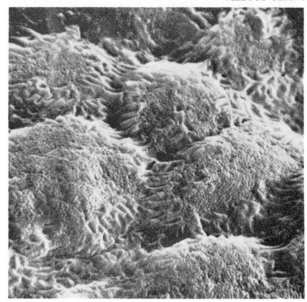

Fig. 9.5   The surface of the jejunum in the rat. The absorptive cells are covered by mucus. The boundaries of individual cells are seen as depressions laced by mucus. The tips of the microvilli are obvious. Scanning electron microscopy. Approximately × 6 330. (Reproduced by permission from Balcerzak, S. P. *et al.* (1970) *Gastroenterology* **58**, 52. © 1970 The Williams & Wilkins Co., Baltimore.)

membrane-bound acidophilic granules that elaborate neutral polysaccharide-protein materials. These chemicals are thought to be synthesized by the rough endoplasmic reticulum on ribosomes, then transported to the Golgi apparatus and become converted into granules which migrate to the cell surface to be secreted. The rough endoplasmic reticulum is well-developed. The nucleus is oval-shaped and basal. Other organelles are shared by any other typical cell.

Paneth cells are thought to arise by differentiation of cells along the sides of the crypts that are intermediate in structure between undifferentiated Paneth cells and goblet cells. Most of these so-called *intermediate cells* move out of the crypts to probably differentiate into goblet cells on the villus; a few remain behind and give rise to Paneth cells. The functions of Paneth cells are uncertain, although current evidence has shown that they may remove ions of heavy metals and secrete amino acids from the body into the intestinal lumen. They may also secrete *lysozyme*, an antibacterial enzyme (also found in human tears) that may regulate the number of bacteria in the intestinal lumen by dissolving their cell wall. They also possibly digest bacteria by phagocytosis and transport bacterial endotoxins across the intestinal epithelium. Paneth cells may secrete enzymes, for example peptidases and lipase, although this seems unlikely.

Small intestinal development

The human small intestine is first a small, straight tube which elongates and con-volutes two months after gestation, when the embryo is 2 cm long. The intestinal surface is initially flat and lined by two to four layers of cuboid cells. When the embryo is about three months old and 8 cm long the intestinal surface becomes single-layered and the villi begin to develop. The surface becomes ridged proximally and finger-like distally. Simultaneously, bud-like diverticulae grow downwards and form rudimentary crypts of Lieberkühn. At first, all the cells along the sides of the villi are capable of dividing, but later cell-division becomes restricted to cells along the sides of the crypts. The first cells to differentiate are the absorptive and goblet cells, then the enterochromaffin cells and finally the Paneth cells. Pinocytotic vesicles and glycogen granules appear in the apical cytoplasm of the developing absorptive cells; the vesicles probably contain materials absorbed from the amniotic fluid. The muscularis externa develops two to three months after gestation and the muscularis mucosa after five months. Brunner's glands appear in the duodenum after four months and Peyer's patches after seven months. By contrast, in rats the development of the villi and that of the cells lining them does not begin until late during gestation and is not completed until after birth.

Villous cell renewal

The lining of the mammlian alimentary canal is continually changing. The small intestinal epithelium is a dynamic, active, proliferating layer which renews itself rapidly and regularly in an orderly manner (Fig. 9.6). Cell proliferation is confined to undifferentiated secretory cells along the sides of the crypts. The birth rate of new cells is about one per 100 crypt cells per hour. These cells possess enzymes necessary for cell division but lack those specialized for digestion. Cells migrate from the sides of the crypts into the vestibules and circumvillar basin before reaching the base of the villus. As the young cells move up the villus and mature, the mechanisms for DNA assembly are abruptly replaced by those for protein synthesis, and their digestive enzymes and absorptive properties progressively increase. So, with increasing age comes increasing efficiency. At the villous tip the mature cells are displaced by others following and are shed into the lumen. The shed cells are responsible for most of the 'free' enzymes in the small intestinal lumen.

D. N. Croft and P. B. Cotton (1973) have estimated that in man up to 50 million cells each minute are lost from the entire normal small intestine. Only about one-tenth of that number is lost from the large intestine and even less from the stomach. The total fresh weight or epithelial cells lost from the small intestine is almost 300 g in 24 hours, representing about 50 g of protein and 12 g of fat, both of which are digested and partly reabsorbed.

The renewal process (turnover time) in the small intestine has been studied by

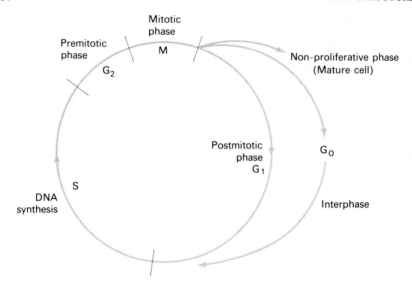

Fig. 9.6   Phases of the renewal cycle of the absorptive cell. Approximate duration M: 1 h, G_1: 10 h, S: 8 h, G_2: 1 h. Mean generation time: 20 h (after Lipkin, 1965). G_0 occupies 95 per cent of the cell cycle.

mitotic counting and by cell labelling, and lasts between one to six days, depending on the species. The rate of epithelial cell renewal varies with age, hormones, the food regime and under various pathological conditions. Once severely damaged, the villous surface is replaced by a protective, flattened, stratified epithelium, incapable of either digestive or absorptive functions.

10　The plasma membrane and its functions

Structure

The plasma (cell) membrane is an ordered structure composed of proteins, lipids and water, serving as a framework for a battery of enzymes that carry out the final stages of digestion and, subsequently, absorption. Several models of the molecular structure of the plasma membrane have been proposed (e.g. Bretscher, 1973; Capaldi, 1974; Finean, 1972; Fox, 1972). None are univerally accepted and it is improbable that there is a unique membrane structure. A currently popular model suggests that the plasma membrane consists of a mobile lipid bilayer with a random array of protruding globular protein molecules, some of them partially embedded in the lipid and others penetrating through it. The main components of the membrane matrix are included in the 'pauci-molecular' model proposed jointly in 1935 by J. Danielli and H. Davson and extended in 1959 by J. D. Robertson as the 'unit (i.e. universal) membrane' model (Fig. 10.1). This classical model has been questioned by several workers during the 1960's but new evidence tends to support the 'unit membrane' concept, although it is undoubtedly oversimplified.

The basic features of the 'unit membrane' model include a protein-phospholipid-phospholipid-protein sandwich, four molecules thick, comprising 60 per cent protein and 40 per cent phospholipid. The phospholipid layers each consist of paired hydrocarbon chains of lecithin (fatty acid) molecules, pointing inwards, which are hydrophobic (water-insoluble), and a terminal group containing choline phosphate (or glycerol) residues, pointing outwards, which are hydrophilic (water-soluble). A molecule of cholesterol is packed in between two molecules of phospholipid. The phospholipid layers are bound firmly together by an outer and an inner coating or protein molecules.

The 'unit membrane' model has received support from transmission electron microscope and X-ray diffraction studies. Viewed under the electron microscope the plasma membrane is found to consist of two dark bands each about 20 Å thick, separated by a central light band, about 35 Å thick (Fig. 9.3). The dark bands each represents a protein monomolecular layer and the terminal choline phosphate (glycerol) group of a phospholipid layer embedded in the protein molecules. The light band represents the mixed phospholipid molecular bilayers.

D. Maestracci and colleagues (1973) have recently separated proteins associated with human jejunal and ileal microvillous membrane by electrophoresis on acrylamide gels, a technique which identifies the number and size distribution of the proteins, protein subunits and glycoproteins. At least 23 protein bands, grouped into two classes, corresponding to a heterogenous group of

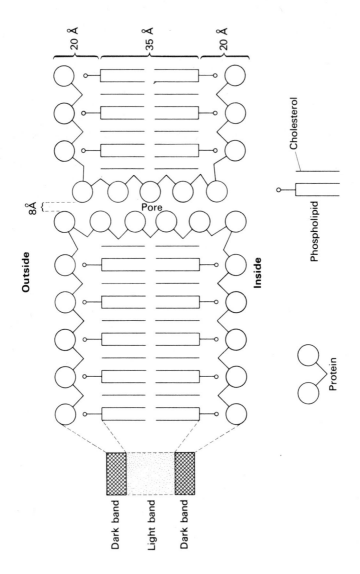

Fig. 10.1 The 'unit membrane' model of the plasma membrane. (Details in text).

polypeptides with molecular weights varying from 23 000 to over 400 000 were isolated. Marked differences were obtained between the jejunum and ileum. About 13 glycoproteins with molecular weights varying from 50 000 to over 400 000 were also identified, some of them probably being associated with maltase, sucrase-isomaltase, enterokinase and alkaline phosphatase, all of which have molecular weights higher than 140 000.

Under the electron microscope pores have not yet been seen in the plasma membrane. However, as water and small lipid-insoluble molecules can easily enter the cell it is probable that the membrane is uniformly pierced by pores of molecular dimensions that follow straight or convoluted paths across the membrane. The pores may be permanent; Danielli (1954) has suggested that such pores are lined by a single layer of protein molecules. Alternatively, the pores may be transient. The surface of the membrane has been likened to a plateful of spaghetti with a marble on top; on shaking the plate the marble eventually finds its way on to the plate surface, yet 'pores' are not seen at any time. Experiments on the diffusion rate of solutes at differing molecular weights into absorptive cells indicate that the pores each have an average diameter of about 8 Å, although other experiments suggest that they are slightly wider. Membrane pores in the human proximal small intestine are twice as large as those distally. Generally, lipid-insoluble substances with a molecular weight of 180 or greater are unable to penetrate the pores.

Membrane pores may have fixed electrical charges lining their walls which will probably affect ion mobility and electrical phenomena. For example, fixed negative charges lining the pores may accelerate cation diffusion through the pores and retard anion diffusion. Thus, membrane pores can discriminate molecules on the basis of size, shape, and charge. Estimations suggest that the microvillous membrane surface may contain 0.4×10^{13} pores per cm^2 (10^6 cells), occupying 0.02 per cent of the total area of the intestinal surface.

The absorption process

The plasma membrane acts both as a selective barrier and a gateway for the passage of different substances into and out of the absorptive cell. There are several ways by which molecules can pass through the cell membrane (Fig. 10.2*):

1 diffusion: (a) through pores, (b) through lipid;
2 carrier-facilitated transport: (a) facilitated diffusion, (b) exchange diffusion, (c) active transport;
3 pinocytosis or phagocytosis.

* Details of the kinetic approach to transport are not given but may be found in several references at the end of this book.

1 Simple diffusion or passive transport

This is a process where the rate of transfer of a non-electrolyte is directly proportional to the concentration gradient across the membrane. This process occurs
from a region of high concentration outside the cell (input side) to a low one inside (output side), and hence substances are not absorbed against a chemical concentration gradient but proceed 'downhill' into the cell. This is the only driving
force of the transport mechanism and since cellular energy is not required,
transport is unaffected by metabolic inhibitors. Also, the sodium molecule is not
required for transport. Some non-electrolytes diffuse through water-filled pores
in the membrane *(passive (pore) diffusion),* while others enter the cell by temporarily becoming soluble and then diffusing across the membrane barrier *(passive
(lipid-soluble) diffusion).* Solutes that enter the cell by the first pathway include
water, some sugar molecules (arabinose, ribose, sorbose), alcohol, drugs and
water-soluble vitamins, while those that enter by the second pathway include
monoglycerides, fatty acids, cholesterol and fat-soluble vitamins. For electrolytes, however, the rate of diffusion depends not only on the chemical concentration gradient but also on the electrical potential gradient across the cell
membrane.

2 Carrier-facilitated transport

Many substances cross the plasma membrane much faster than can be accounted
for by simple diffusion. For this reason, it is postulated that molecules cross the
membrane temporarily adhering to a mobile *carrier.* Each carrier has two
specific binding sites and can handle two molecules simultaneously. (Other
carrier models have been proposed.) When loaded, therefore, the carrier consists
of three molecules; such a carrier is said to be a *ternary complex.* The carrier
picks up two molecules (usually sodium and non-electrolyte) at the outer surface
of the membrane then, loaded, it moves across the membrane by either rotation
or diffusion to deposit the two molecules on the inner membrane surface inside
the cell, so achieving intracellular concentration. Thus, solute and sodium enter
the cell in a molecular ratio of 1:1. The empty carrier then shuttles back across
the membrane and picks up two more molecules. It is possible, however, that
once sodium and solute have been discharged inside the cell the carrier can pick
up potassium at the empty sodium site and remove it from the cell down its concentration gradient, so affecting a two-way exchange between sodium and
potassium. The carriers can move to and fro very quickly; estimations suggest
that a carrier can do so about 180 times each second.

Little is known about the nature of the carrier. In 1974, R. G. Faust and S. J.
Shearin were the first to isolate a D-glucose and L-histidine homogeneous protein carrier, with a molecular weight of about 55 000, from the microvilli of
hamster jejunum (carrier proteins with a M.W. of 32 000 involved in sugar and
amino acid transport ('permeases') were earlier isolated in bacterial membranes;
it is estimated that a bacterium has 10 000 such carriers). The binding sites are

highly specific to certain molecules and can become blocked by other molecules and poisoned or denatured by chemical reagents. Estimations propose that there are 26×10^5 carriers per absorptive cell surface, or 1700 carriers per μm^2 membrane surface (other estimates indicate that fewer carriers are available). Assuming that each carrier is 100 Å in diameter, carriers would then occupy about 12 per cent of the total membrane surface area. Hence, the surface ratio of carriers to pores in the apical membrane is about 600 : 1.

Although the significance of the 'unstirred' water layer overlying the plasma membrane has not yet been assessed, it is possible that this surface may significantly alter the parameters of both simple diffusion and carrier-mediated transport. Moreover, 'bound' water associated with the lipid and protein components of the plasma membrane is also likely to modify absorptive processes.

(a) FACILITATED DIFFUSION This type of transport across the plasma membrane has features common to both simple diffusion and active transport (below). It is still a diffusion process but facilitated by a membrane carrier. Like simple diffusion, sodium ions are not essential for the transport process and some solutes will be transferred in the presence of either potassium or lithium ions. Moreover, diffusion proceeds 'downhill' along a concentration gradient and therefore does not require any expenditure of cellular energy. Also, net solute movement is from a higher to a lower electrochemical gradient. Like active transport, it is subject to substrate specificity, competitive inhibition and saturation kinetics. Most solutes probably leave the base of the absorptive cell by this process. Facilitated diffusion has been chiefly studied in relation to the red blood cell, where glucose is transported inwards 1000 times faster than can be accounted for by simple diffusion. Estimations have shown that for each red blood cell there are 700 000 transport sites.

(b) EXCHANGE DIFFUSION (COUNTER-TRANSPORT) The transport of some amino acids shows this curious phenomenon, which has many features in common with facilitated diffusion. But for each molecule transported by a carrier across the membrane into the cell, an analogous molecule is transported by the same carrier in the opposite direction out of the cell. However, this system ultimately produces a net uptake of solute. The purpose of the bi-directional flux of structurally analogous molecules is obscure.

(c) ACTIVE TRANSPORT Many sugars and amino acids are transported across the plasma membrane by active transport. This operation involves the movement of solutes from a region of low concentration outside the cell to one of high concentration inside the cell, that is, movement occurs 'uphill' against both chemical concentration and electrochemical gradients.

In 1933 T. Wilbrandt with L. Laszt and E. Lundsgaard independently suggested that glucose was first phosphorylated at the apical membrane of the absorptive cell, then transferred across the cell and dephosphorylated, releasing

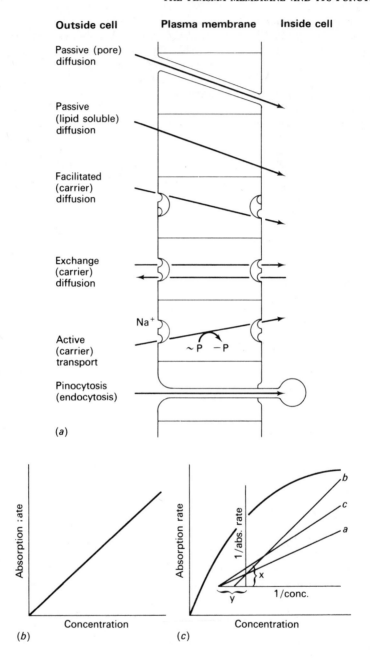

(a)

(b)

(c)

glucose which diffused through the basal membrane *(phosphorylation theory)*. This concept has now been proved to be incorrect. Although E. W. Reid in 1902 first speculated that sodium stimulated glucose absorption, it was not until the seven-year period between 1958–1964 that several workers showed that the sodium molecule was essential for active transport and in its absence many solutes cannot be transferred into the cell. In 1962 R. K. Crane and co-workers first suggested that the sodium molecule crossed the apical membrane with the solute as a carrier complex. Both solute (glucose) and sodium were released inside the cell, but the sodium molecule was then removed from the same surface and passed back into the lumen.

However, further investigations showed that the drug ouabain in the serosal but not mucosal fluid (p. 90) inhibits the outward movement of sodium molecules (ouabain, a heart stimulant, prevents the action of membrane ATPase), while the chemical phloridzin (phlorrhizin) blocks the coupled sodium-solute (glucose) entry at the mucosal (apical) but not serosal (basal) surface (phloridzin competes with glucose for a common carrier binding site*). These observations led S. G. Schultz and R. Zalusky in 1964 to suggest that coupled sodium and solute entered the cell through the apical membrane as a phloridzin-sensitive ternary complex, and that the sodium molecule was actively pumped out by an ouabain-sensitive, active sodium transport mechanism at the basal membrane. The sodium pump uses metabolic energy to create an electrochemical potential

---

* A microvillous enzyme, phloridzin hydrolase, transforms phloridzin to phloretin at the site of interaction. Hence, phloretin and not phloridzin is possibly the inhibitor of sugar transport.

Fig. 10.2   (a) Types of transport mechanisms across the plasma membrane. Slope of arrows indicate non-electrolyte gradients (adapted from Pike and Brown, 1967).

(b) The rate of solute absorption by simple diffusion is directly proportional to its initial concentration in the intestinal lumen (Fick's Law).

(c) Kinetics of intestinal transport (p. 83). For an actively transported solute, the rate of transport is not proportional to its initial concentration but approaches a limiting value with increasing concentration (Michaelis-Menten kinetics). The reciprocal of the absorption rate plotted against the reciprocal of the initial concentration (*inset*) gives a straight line (Lineweaver-Burk plot). The reciprocal of the distance x from the origin indicates the limiting velocity of transport (V_{max}), and the reciprocal of the distance y indicates the concentration at which the absorption rate is half-maximal (K_t). A low K_t value indicates that the transport mechanism easily becomes saturated with the solute, or that it has a high affinity for the transport mechanism. A high K_t has a low affinity.

Such values may be used for investigating interactions of different solutes on active transport in absorptive cells. With *competitive inhibition* the presence of one solute (the competitor) diminishes the ease of access of another solute to the carrier site. Here, x will remain unchanged and y will decrease (e.g. '*a*' no inhibitor; '*b*' with inhibitor). With *non-competitive inhibition* the carrier site is unaffected but the inhibitor interferes with cellular production or utilization of energy. Here, x will alter and y will remain unchanged (e.g. '*a*' no inhibitor; '*c*' with inhibitor) (p. 123).

difference (p. 92) which in turn supplies energy to activate the carrier and dissociate the carrier complex inside the cell. Accumulated solutes inside the cell may be partly metabolized by the cell before crossing the basal membrane, probably by facilitated diffusion.

This working model (Fig. 10.3) of the absorptive cell is at present generally accepted as being the most plausible one for the active transport of many solutes. By isolating the apical membrane from the lateral-basal membranes recent studies have confirmed that sodium is essential for solute (glucose) entry whereas

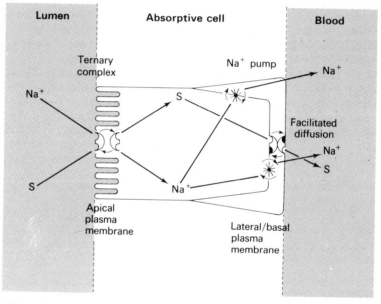

Fig. 10.3   Mechanism for active transport in the absorptive cell (slope of arrow indicates gradient). S: solute. Details in text (modified from Schultz and Curran, 1970).

it is not required for solute exit. However, histochemical and autoradiographic evidence have shown that membrane ATPase, an enzyme intimately related with the sodium pump, is mainly located in the lateral plasma membrane. Thus, the sodium pump is more likely to be situated along the lateral membrane border of the intercellular spaces than along the basal membrane.

Finally, the effect of ouabain on the sodium pump mechanism is species specific. While ouabain in the serosal fluid virtually stops the sodium pump in the rabbit or guinea-pig intestine, it barely affects the mechanism in the rat or mouse intestine. Hence, either an ouabain-insensitive or an additional, as yet unidentified, enzyme system must be postulated to account for sodium extrusion in the small intestine of these species. However, such conclusions are possibly unwarranted for the techniques used to assess the effects of ouabain on sodium transfer prevented the full effects of the drug acting on the sodium pump.

Another mechanism by which sodium ions can regulate the absorption of solutes in the (human) small intestine has been proposed by J. S. Fordtran and colleagues (1968), a process termed *solvent drag*. Free solute (glucose) passes through the microvillous membrane and enters the intercellular spaces, creating an osmotic pressure gradient for bulk water movement, which filters through the membrane pores and/or the sieve-like tight junctions, sweeping sodium chloride in the stream of water into the intercellular channels. (Aqueous pores, however, have not been seen along the tight junctions. In fact, freeze-etched preparations indicate that the junctions resemble a modified zipp with interlocking units making head-to-head contact.) This model assumes that solute and sodium are not linked by a common membrane carrier.

The importance of sodium ions in active transport has been disputed. Several workers have recently found that glucose or galactose absorption, particularly with *in vivo* studies, remained little affected in the absence of sodium in the lumen, indicating that these ions are apparently not essential for the absorption of either sugar. However, these observations do not necessarily disprove the sodium gradient hypothesis. It is possible that the 'unstirred' water layer, over and amongst the filaments of the glycocalyx, constitutes a diffusion barrier for sodium secreted by the absorptive cells in the absence of sodium in the intestinal lumen, so creating a local, high concentration of sodium molecules for the coupled solute-sodium movement through the apical plasma membrane. Moreover, an additional but weak sodium pump has recently been found at the apical border which, at low sodium luminal concentrations, serves to expel intracellular sodium into the lumen for subsequent sodium-solute coupling at the microvillous membrane. However, further experiments are necessary to confirm the validity of this second pump mechanism.

The proposed model (Fig. 10.3) of the absorptive cell contrasts with that of many other mammalian cells as, for example, the muscle cell (Fig. 10.4). In both, the intracellular concentration of sodium and chloride ions is low and that of potassium high. Outside the cell, the sodium and chloride concentrations are high and the potassium concentration low. In both, an electrical potential difference is maintained across the cell. In the muscle cell, however, the exchanges of sodium and potassium between the cell interior and its exterior are coupled by a sodium pump located in the membrane surrounding the cell which is driven by metabolic energy (p. 13). By contrast, in the absorptive cell sodium and solute enter one surface on a linked carrier. Sodium is then pumped out by an energy-requiring mechanism at the opposite (or lateral) surface while the solute diffuses through that surface. The absorptive cell can thus be considered asymmetrical and the muscle cell symmetrical. The transport of potassium in and out of the absorptive cell is probably a passive process, independent to that of

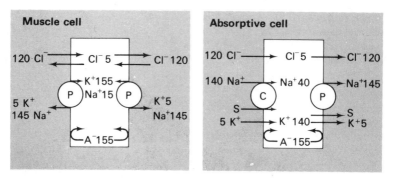

Fig. 10.4   Transport of electrolytes across a muscle cell and an absorptive cell. P: pump; C: carrier; S: solute. Value of electrolytes in mmol l^{-1} (details in text).

sodium. The mechanism for chloride entry and exit in both cell models is uncertain.

Energy requirements for entry

Investigators all agree (but see above) that sodium is required for solute entry but disagree over the form of energy required for sodium-solute transport across the apical membrane of the absorptive cell. R. K. Crane (1962, 1965) suggested that the potential energy is mainly derived from the inwardly-directed sodium gradient, which is maintained by the sodium pump at the basal membrane, thus allowing a low intracellular sodium concentration. Recent work indicates that nearly three-quarters of the total intracellular sodium exists in bound form, and hence the effective free sodium concentration (activity) gradient is much greater than that actually measured. T. Z. Csaky (1963) proposed that the process also required metabolic energy (ATP). Probably only about ten per cent of the total energy available in the absorptive cell is required for active transport. More recently, other workers have postulated that an enzyme system is involved. G. A. Kimmich (1970) showed that when the normal intracellular sodium gradient was reversed, solutes (galactose) were nevertheless transported into the absorptive cell at the same rate. Absorptive cells were first pre-loaded with sodium and galactose in the cold and then incubated in a warm medium, containing galactose and a low sodium concentration, when they unexpectedly accumulated the solute although the intracellular concentration of sodium was higher than that of the medium. This is inconsistent with the sodium gradient hypothesis when an active extrusion of the solute was predicted. Kimmich postulated that the energy necessary for inward solute movement is driven by transport or membrane sodium-potassium activated (apical) adenosinetriphosphatase (membrane ATPase), which generates two phosphorylated intermediates (acyl phosphates)

$(E_1 \sim P$ and $E_2 \sim P)$ during catalysis of ATP hydrolysis. The reaction involves sodium, potassium and magnesium ions:

$$
\begin{array}{l}
\text{E + ATP} \underset{}{\overset{Na^+, Mg^{2+}}{\rightleftharpoons}} E_1 \sim P + ADP \\
\text{(free} \\
\text{enzyme)} \qquad\qquad\qquad\qquad Mg^{2+} \\
\qquad\qquad\qquad K^+ \\
\text{E + P}_i \rightleftharpoons E_2 \sim P + H_2O \\
\text{+ non-electrolyte} \\
\text{transport}
\end{array}
\left.\vphantom{\begin{array}{c} a \\ b \\ c \\ d \end{array}}\right\} \begin{array}{l} \text{Phosphorylated} \\ \text{intermediates} \end{array}
$$

Thus, Kimmich believes that the energy for the hydrolysis of the reaction can provide the energy necessary for transfer at the luminal side of non-electrolytes, for example, sugars or amino acids. Different workers have found that Mg, ATPase activity is highest in the adult rat proximal small intestine, whereas Na, K, ATPase activity is highest in the mid-intestine and lowest in the distal intestine, suggesting that the activity of the second enzyme is involved in the active transport of some nutrients (monosaccharies) which are absorbed mainly in the mid-intestine. Other investigators have found a direct correlation between transport of amino acids in Ehrlich ascites tumor cells or mouse pancreas cells and cellular metabolism, but not with the magnitude of the sodium gradient.

Although membrane ATPase was originally isolated both in the mucosal (apical) and serosal (basal) membranes of the absorptive cell, more refined techniques have shown that this enzyme is in fact mainly concentrated in the lateral plasma membrane, probably at the inner-protein face (p. 88). Hence, the importance of (apical) membrane ATPase in generating the inward sodium-solute movement by hydrolysis of cellular ATP appears questionable. Despite this, Na, K, ATPase has recently been isolated from the microvillous border, with higher activity at the villous tip than in the crypts. But other evidence has thrown doubt on the importance of cellular metabolism in the active transport process. H. J. Leese and J. R. Bronk (1972) incubated slices of mucosa from rat jejunum, either in the presence or absence of glucose. In both experiments the ATP content of the tissue declined, indicating that glucose uptake by the absorptive cells is independent of ATP hydrolysis. These observations suggest that the driving force for glucose entry is the downward sodium gradient from the exterior to the interior of the cell, in agreement with the sodium gradient hypothesis. Perhaps the energy released from cleavage of superficial dissaccharidases and tripeptidases by hydrolases might be available for carrier movements deeper in the membrane.

H. Murer and U. Hopfer (1974) have recently proposed an alternative inward driving source. They provided evidence to suggest that the main driving impetus for inward sodium-solute (glucose) movement might reside in the electrical potential difference (p. 95) between the inside of the absorptive cell and the outside. These workers postulate that electropositive sodium ions in the loaded ternary complex are attracted across the apical membrane to the negative cell interior, then sodium and solute are released intracellularly.

Finally, several reports currently suggest that transport by the small intestine may be partly under the control of the autonomic nervous system. For example, P. T. Hardcastle and J. Eggenton (1973) found that acetylcholine, a neurohumoral transmitter, in the serosal fluid (p. 96) of the rat jejunum stimulated an electrical potential difference, presumably by diffusing through the submucosal layers before reaching cholinergic receptors in the viscinity of the epithelium. Thus, acetylcholine released form the nerve endings may regulate the electrogenic properties of the epithelial cells, resulting in intestinal absorptive or secretory activities.

Owing to the conflicting evidence currently available it is not possible to state precisely what form of energy is required for non-electrolyte entry into absorptive cells. It is quite likely that several mechanisms, working in unison, are necessary for optimal transport.

3 Pinocytosis or phagocytosis (endocytosis)

Although pinocytosis means 'cell drinking' and phagocytosis means 'cell eating', in absorption studies the term pinocytosis is generally applied to both methods of transport (the term 'endocytosis' seems more appropriate). Pinocytosis is a process involved in the selective uptake of large molecules from outside the cell into the cell interior. Pinocytosis starts by binding large molecules to the outer surface of the cell membrane. The membrane then sinks into the cell, forming a pinocytotic tube with the enclosed molecules within a membrane-bound vesicle at the base of the tube. The vesicle then detaches from the tube, leaving the material inside the cell, enveloped by the membrane. The membrane eventually dissolves, releasing the molecule inside the cell. Any fluid in the vesicle is presumably expelled by diffusion, and intracellular concentration of the molecules is thus achieved. Pinocytotic transport can proceed equally both in and out of the cell and the direction of transport depends upon the concentration gradient and luminal flow, that is, by a process similar to passive transport. However, the fact that pinocytosis can be inhibited by factors that interfere with metabolism (poisons, low temperature), and exhibits saturation kinetics, suggest that transport by this process may have some features in common with active transport. Pinocytotic vesicles participate in immunoglobulin uptake in the suckling intestine and in the transfer of digested fat molecules from the absorptive cell into the circulation.

Electrical changes in relation to intestinal transport

In many mammalian cells the plasma membrane maintains a difference in both the ionic composition and electrical potential between the internal and external environment, the inside of the cell being negative in respect to the outside. The mechanism involves exchanges of potassium outside the cell with sodium inside the cell and postulates a *linked* or *coupled pump* in the membrane (p. 16). In some epithelial cells, including the absorptive cells, differences between the functions of the apical and basal membranes result in both the net movement of molecules (that is, absorption) and an electrical potential difference across the cells. The active transport of solutes proceeds 'uphill' against an electrochemical gradient, unlike any other form of transport which occurs in the same direction as the electrochemical gradient.

In the absence of sodium ions in the lumen of the rat distal small intestine, or when the level of sodium is reduced and replaced by impermeant ions (e.g. Tris),

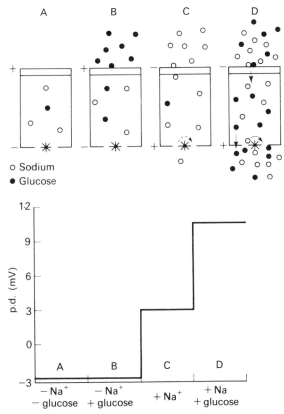

Fig. 10.5   Electrical potential difference (p.d.) across the (rat) absorptive cell in (a) the absence of both sodium and glucose, (B) the absence of sodium and presence of glucose, (C) the presence of sodium, and (D) the presence of both sodium and glucose.

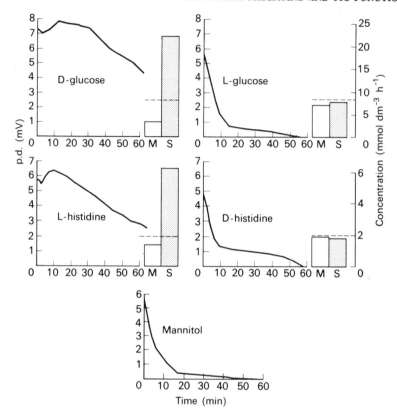

Fig. 10.6   Electrical potential difference (p.d.) at 5 min. intervals for 1 h at 37°C in a 95 per cent O_2 +5 per cent CO_2 mixture, using whole everted *in vitro* mouse small intestine. D-glucose and L-histidine are actively absorbed while L-glucose and D-histidine are not absorbed. Initial concentration of solutes shown as broken lines; M: final mucosal concentration, S: final serosal concentration. The relatively high p.d. for D-glucose or L-histidine is associated with solute-sodium entry and the electrogenic sodium pump. The low p.d. for L-glucose or D-histidine is probably associated with the endogenous sodium pump. Mannitol (2 or 10 mmol dm^{-3}), which is neither absorbed nor metabolized, was used as a control. (Original).

there is a small electrical potential difference of at least −3 mV across the absorptive cell, the mucosal surface being positive and the serosal surface negative (Fig. 10.5). The presence of sodium ions causes the electrical potential difference to revert to about +3 mV, the mucosal surface becoming negative and the serosal surface positive (Fig. 10.5). Since external energy is not supplied this biological pump is called an *endogenous sodium pump*.

When glucose is added to the luminal solution in the presence of sodium the electrical potential difference immediately rises and remains at about $+10$ mV (Fig. 10.5). The magnitude of the potential depends on the glucose concentration. The presence of phloridzin on the mucosal side prevents the entry of glucose by competitive inhibition and abolishes the electrochemical gradient, but the chemical has little effect on the serosal side. Ouabain on the serosal side blocks the sodium pump and also abolishes the electrochemical gradient, but it has little effect on the mucosal side. Thus, the increased electrical potential difference is related both to the influx of sodium ions when coupled glucose and sodium enter the cell apex and to the efflux that occurs when sodium ions are pumped from the base or sides of the cell. Active transfer of many non-electrolytes increases the potential owing to a rise in the serosal potential while the mucosal potential remains virtually unchanged. This pump is referred to as an *electrogenic (rheogenic) sodium pump*. Results of experiments showing sodium pump action in relation to solute absorption are given in Fig. 10.6.

Another pump, a *non-electrogenic (neutral) sodium pump,* has also been identified. (It is possible, however, that this pump is an endogenous sodium pump.) The entry of some solutes into the absorptive cell causes increased sodium movements but they do not generate an electrochemical gradient. Mannose, which is not transported at the mucosal surface, becomes metabolized after entering the absorptive cells at the serosal surface and operates such a pumping mechanism. Unfortunately, the situation is more complicated than this. Fructose, like glucose, which is both actively transported and metabolized in the rat small intestine, promotes non-electrogenic transport of sodium. Galactose, which is actively transported but not metabolized, promotes electrogenic transport yet does not stimulate sodium transfer. Clearly, the interrelationships between the factors involved in electrical phenomena of absorptive cells are still incompletely understood.

11 Methods

Although attempts to investigate the absorption of food and water by the small intestine were first made over 350 years ago, until recently very little was known about the processes involved in absorption. Revival of interest in intestinal absorption was initiated in 1954 by T. H. Wilson and G. Wiseman who popularized the everted sac technique. Since then probably nine investigations in ten use this method, and as a result rapid progress is being made in elucidating fundamental mechanisms on intestinal absorption and transport.

In vitro methods

Undoubtedly the reason for the popularity of the everted sac technique lies in its simplicity. As D. H. Smyth (1967) aptly put it: "There is a fundamental law governing progress in the experimental sciences, that if one thinks of something easy to do, which has not been done before, further examination shows that it isn't easy or that it has been done before! Periodically the introduction of some new technique grants temporary relaxation of this law, and the introduction of the isolated intestine is a good example of this". Briefly, this is how an isolated intestinal preparation can be made (Fig. 11.1). The small intestine of a small laboratory mammal is quickly removed under anaesthesia and the intestinal lumen flushed out with 0.9 per cent saline solution to clean it of adhering food debris. A metal or glass rod is gently pushed through the lumen, the tip tied to one end of the intestine and the whole preparation carefully turned inside out by pulling the rod backwards. The everted intestine is then separated from the rod, ligatured at one end, and filled with the test substance dissolved in oxygenated Krebs Ringer saline-bicarbonate buffer solution by a syringe via a blunt needle inserted into the open end. This end is then ligatured and the sac put into the same test solution in an incubating flask. The flask is first gassed for a few minutes with a mixture of 95 per cent oxygen and five per cent carbon dioxide, then sealed and mechanically shaken in a water-bath at a temperature of 37°C for not more than one hour. The solution inside the sac is then removed and the change in concentration of the solute inside and outside determined by chemical analysis. Radioactive tracers are increasingly being used to measure absorption.

The incubation fluid is referred to as the *mucosal fluid* and the fluid inside the sac as the *serosal fluid*. If the sac is weighed empty and then filled, both before and after the experiment, the *serosal fluid transfer* and *gut fluid uptake* can also be found by difference in weight. The preparation can also be homogenized in

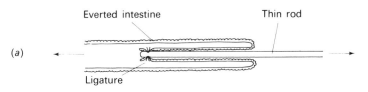

(a)

Everted intestine   Thin rod

Ligature

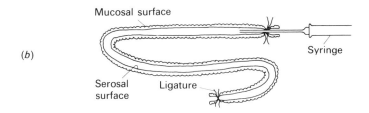

(b)

Mucosal surface

Syringe

Serosal   Ligature
surface

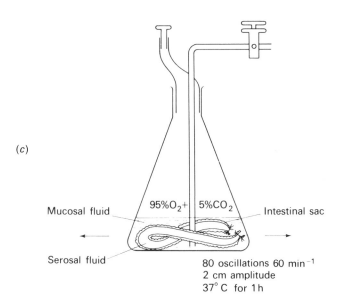

(c)

Mucosal fluid   95%O$_2$+ 5%CO$_2$ Intestinal sac

Serosal fluid   80 oscillations 60 min^{-1}
                       2 cm amplitude
                       37° C for 1 h

Fig. 11.1 *In vitro* method for preparing an everted small intestinal sac. (*a*) The intestine is turned inside out by using a thin rod. (*b*) The everted intestine is filled with serosal fluid by means of a blunt needle fitted to a syringe. (*c*) The sac is incubated in the mucosal fluid in equilibrium with a mixture of 95 per cent O$_2$ + 5 per cent CO$_2$ and shaken in a water-bath at 37°C for up to 1 h.

distilled water and the amount of solute taken up by the tissues determined in the supernatant.

Other *in vitro* methods are also currently in use. An open-end everted sac can be suspended in the mucosal fluid in a heated test tube and serosal samples taken at regular intervals during the experiment. The test fluid can be continually *circulated* by two independent systems, both inside and outside an isolated intestine, and the rate of transfer of the solute then found. This method is especially useful for intestinal preparations with large animals. An everted intestine can be cut into *small rings,* the rings incubated in the test solution and the amount of solute accumulated by the tissue during incubation then determined by homogenizing the tissue and analyzing the supernatant. The inner surface of the intestine can also easily be *stripped* away from its underlying supporting tissue, the mucosal strip stretched across a central partition of a chamber filled on both sides with the test solution and the rate of solute transfer across the mucosal strip determined at intervals. Villi can be *isolated* from the rest of the intestine by putting intestinal pieces into a syringe and forcing the tissue through needles of successively smaller size, then collecting the separated villi. Finally, *individual absorptive cells* can be separated mechanically, chemically or enzymatically from a mucosal strip and then incubated.

Care must be taken when expressing the results. Providing that the intestine is healthy, the results can be expressed in absolute amounts of solute gained or lost per unit time on a fresh weight basis. However, under certain conditions, for example, when the intestine is diseased, the fresh weight may change compared with a healthy intestine, and the results are then more accurately expressed as an oven dry weight or on a protein content basis. Alternatively, transfer is sometimes expressed as a final serosal to mucosal concentration ratio or, for tissue preparations, as a final tissue to mucosal concentration ratio.

In vivo methods

In vivo methods are those in which the small intestine is left *in situ* with an intact blood supply. One of the most frequently used methods is to expose the intestine of an anaesthetized animal by a mid-line body incision and to *perfuse* the intestine with the test substance dissolved in saline-bicarbonate buffer solution, continuously oxygenating the circulating fluid at 37°C (Fig. 11.2). Small samples of the test solution can be withdrawn at intervals and the rate of absorption by the intestine then measured. The amount of fluid absorbed in the same preparation can also be determined by using a non-absorbable *marker dye* (for example, pherol red) in the circulating fluid and measuring the increase in concentration of the dye as the fluid is absorbed. Alternatively, the intact intestine can be ligatured at both ends, the test solution *injected* into the loop, the solution recovered at the end of the experiment and the amount of solute that is absorbed then found. Also, the amount of the test substance appearing in the *mesenteric veins* can be determined.

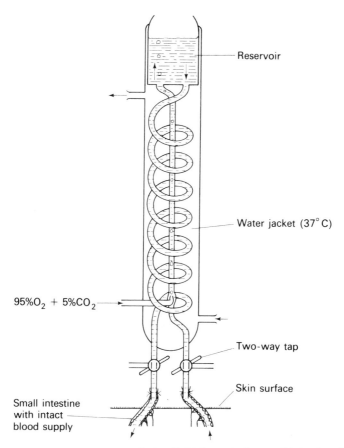

Fig. 11.2   *In vivo* method for perfusing fluid through the small intestine. The fluid is continuously circulated by means of a gas-lift. Samples are withdrawn from the open top for determining solute absorption. The intestinal lumen can be rinsed by two-way taps (modified from Sheff and Smyth, 1955).

Experiments on fully conscious animals can be made. An animal can first be given the test solution by *mouth,* the animal killed after a definite period, and various regions of the alimentary canal, or portions of the small intestine, then quickly ligatured and the amount of test solute remaining in the lumen of the different regions found. Also, the animal can simply be given the test substance in a meal and the amount recovered in the *faeces* or *urine* estimated. Alternatively, a *marker substance,* for example chromium sexquioxide, can be given in the diet containing the test substance and the ratio of the marker and test substance determined both in the diet and in the faeces.

Absorption studies with large mammals sometimes utilize a *Thiry-Vella loop* preparation. The small intestine is first transected at two points a short distance apart. The main intestinal portion is then surgically joined, and the isolated portion, with intact blood supply, made into a loop and joined around two separate incisions in the skin surface. The test solution in the intestinal loop can be continually circulated if required, or two loops can be prepared and connected externally to form a double re-entrant system.

Finally, clinical investigations in man frequently use *intubation techniques*. A thin, pliable triple-lumen tube is passed through the mouth and stomach and into the lumen of the jejunum. The test solution is pumped in through the first tube at a constant rate which perfuses the jejunum, the jejunal contents aspirated through the second tube, and the absorption rate of the solute in the aspirate then found. Contamination is avoided by connecting a small inflatable balloon to the third tube, thus occluding the intestinal lumen below the two open-ended tubes. A reference compound, for example polyethelene glycol, which is neither absorbed nor metabolized, can be perfused simultaneously to indicate changes in the volume of fluid during its passage through the jejunum.

Comparisons between in vivo and in vitro methods

During absorption various substances are either transported or diffuse through the microvillous membrane into the absorptive cell, a process involving different pathways, and they leave the cell through the basal or lateral membranes to enter the fluid in the sub-epithelial tissues, from where they enter the blood capillaries or lacteals and are carried away from the intestine.

In vivo and *in vitro* conditions obviously differ, and it is important to remember this fact when interpreting the results (Fig. 11.3). Under *in vivo* conditions unlimited oxygen and energy are provided to the tissues. Oxygen is supplied by branches of the abdominal aorta, and energy is supplied by nutrients (mainly glucose) in the bloodstream, nutrients in the intestinal lumen and nutrients stored in the absorptive cells. The energy is used both for solute transfer and for maintaining the tissue in a physiologically active state. Moreover, solutes are rapidly removed in the circulation so that they do not accumulate in the intestinal tissues.

Under *in vitro* conditions (everted sac technique), dissolved oxygen is supplied in the mucosal and serosal fluids; energy is available stored in the cells and, if present, in the mucosal fluid. Under these conditions both oxygen and nutrients are available for only a limited period. Also, the substances transferred from the cells cannot be transported away and will accumulate in the serosal fluid, probably by two routes, by the torn capillaries and by diffusion through the different tissue layers of the intestine (Fig. 11.3). Moreover, there is a possibility of some back-diffusion and accumulation of solutes in the tissues which may lead to erroneous interpretations of the results. Finally, everting the intestine may alter its permeability characteristics and damage its structure. Despite these limitations,

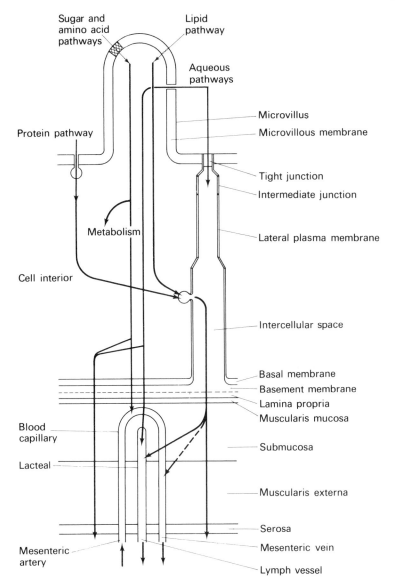

Fig. 11.3 Different pathways in the small intestine available to various solutes under *in vivo* and *in vitro* conditions. With *in vitro* preparations the products of transfer appear in the serosal fluid instead of the bloodstream.

however, the *in vitro* everted intestinal sac preparation remains a valuable asset in physiological investigations. The preparation is simple and quick to prepare and easy to manipulate, absorption in different regions of the intestine in an animal can be studied and fluid transfer easily determined by weighing.

12 Carbohydrates

Carbohydrates (Table 12.1) serve as an important source of energy for vital activities, providing about half the total calorie intake in the Western diet and up to four-fifths elsewhere. Two-thirds of the ingested carbohydrates are derived from starch, one-quarter from sucrose and the remainder includes small amounts of lactose, fructose and glycogen. These substances (except for fructose) are not absorbed by the small intestine as such but are all hydrolyzed to monosaccharides during the absorption process.

Extracellular digestion

Starch is first digested in the mouth and lumen of the proximal small intestine. Starch consists of 80 per cent amylopectin and 20 per cent amylose. Amylopectin consists of straight chains of about 25 glucose units linked at the 1,4-α position and 1,6-α linked branching chains. Amylose consists of straight chains of 250 to 300 1,4-α linked glucose molecules.

Salivary and, mainly, pancreatic α-amylase rapidly hydrolyze the 1,4-α links of both amylopectin and amylose molecules, releasing maltose, maltotriose and isomaltose (α-limit dextrins) (Fig. 12.1). Maltose and maltotriose are 2- and 3-unit pieces of straight glucose chains, and isomaltose consists of intact 1,6-α branch points with short 1,4-α linked glucose chains attached. These disaccharides, together with others in the diet, accumulate in the small intestine during luminal digestion and cannot be absorbed as such. Very little glucose is released.

Membrane digestion

In 1961 D. Miller and R. K. Crane were the first to isolate several disaccharidases from intact microvilli of the hamster small intestine. Their results have subsequently been confirmed by other workers using different animals. The enzymes are localized within the outer layer of the microvillous membrane, and their function is to hydrolyze disaccharides within the membrane into their constituent monosaccharides (Fig. 12.2). Owing to their large size α-amylase molecules M.W. 45 000) cannot penetrate the absorptive cell membrane. However, intracellular starch digestion is apparently possible in the absence of salivary and pancreatic α-amylase, for five different types of intracellular amylases have recently been identified. Therefore, intracellular amylases may complete the hydrolysis of starch begun by salivary and pancreatic α-amylase.

Table 12.1 Classification of carbohydrates

I *Polysaccharides*	
1 Pentosans $(C_5H_8O_4)$ x	e.g. xylan
2 Hexosans $(C_6H_{10}O_5)$ x	e.g. Starch, dextrin, glycogen, inulin
3 Mixed polysaccharides	e.g. hemicellulose, pectin, mucilage
II *Trisaccharides*	
$C_{18}H_{32}O_{16}$	e.g. raffinose
III *Disaccharides*	
$C_{12}H_{22}O_{11}$	e.g. maltose, sucrose, lactose
IV *Monosaccharides*	
1 Pentoses $C_5H_{10}O_5$	e.g. xylose, arabinose, ribose
2 Hexoses $C_6H_{12}O_6$	
(a) aldohexoses	e.g. glucose, galactose, mannose
(b) ketohexoses	e.g. fructose

Fig. 12.1 Digestion of starch (G: glucose units) (adapted from Holmes in Dawson, 1971).

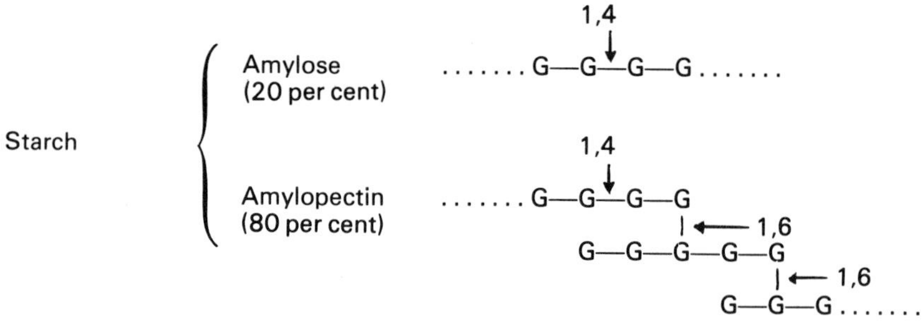
Starch
Amylose
(20 per cent)
1,4
.......G—G—G—G.......
Amylopectin
(80 per cent)
1,4
.......G—G—G—G
1,6
G—G—G—G—G
1,6
G—G—G.......

Salivary and pancreatic α-amylase act on the 1,4-α links only, resulting in

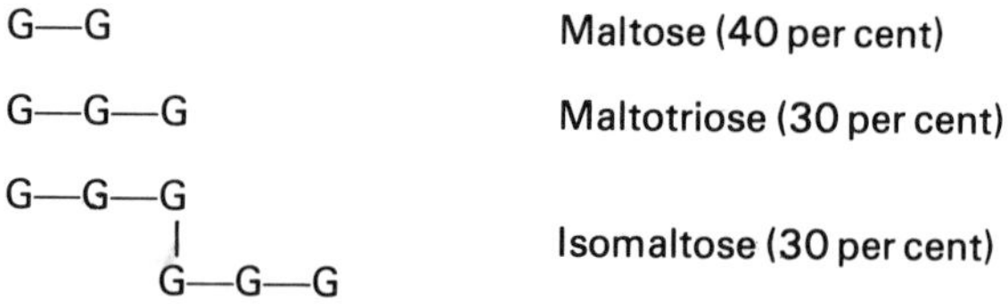
G—G
Maltose (40 per cent)
G—G—G
Maltotriose (30 per cent)
G—G—G
G—G—G
Isomaltose (30 per cent)

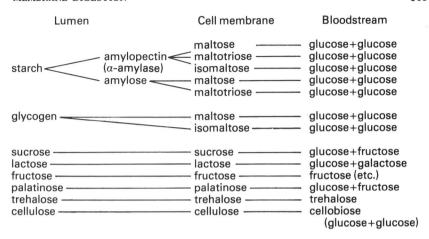

Fig. 12.2   Digestion of carbohydrates in the small intestine.

Disaccharidases show differing *relative activities*. In several monogastric animals the proportion of maltase, isomaltase (palatinase), sucrase, lactase and cellobiase in the microvillous membrane is in the ratio 6:2:2:1:0.1 respectively. The enzymes are usually concentrated in the jejunum, where maximal monosaccharide absorption occurs, at least in some species. Very little hydrolytic activity has been found in the small intestinal secretions or in the epithelium of the stomach, caecum and large intestine. However, there is significant lactase activity of bacterial origin in the lumen of the caecum. Ruminants lack both sucrase and palatinase since fructose, a hydrolytic product of sucrose and palatinose, is not utilized in their diet. Heat inactivation and chromatographic studies have separated at least four different maltases, two different sucrases and two different lactases in the human small intestinal mucosa. The activation of some disaccharidases (for example, sucrase) probably depends on the optimal concentration of sodium in the intestinal lumen. However, some transfer of the hydrolytic products of disaccharides occurs in the absence of sodium ions in the lumen. Disaccharidases (and dipeptidases) can also cause electrogenic sodium pump action resulting from the transfer of hydrolytic products.

The pattern of disaccharidase *development* is paralleled by changes in the diet in growing animals. In young rats the activity of intracellular lactase, which hydrolyzes lactose in milk, remains high until weaning and then declines, while the activities of other disaccharidases associated with a solid diet rise just before weaning and thereafter remain high (Fig. 12.3). Once the adult pattern of disaccharidases is established changes in dietary carbohydrates produce only small quantitative changes in these enzymes. By contrast, in man the adult levels of disaccharidases are attained before birth. The *stimulus* that increases the activities

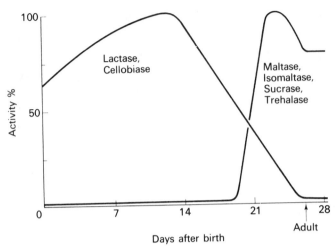

Fig. 12.3 Disaccharidase activities in the small intestine of suckling and adult rats (adapted from Rubino *et al.*, 1964).

of intestinal disaccharidases during development is uncertain, although there is evidence in rats to suggest that hormones of the pituitary (adrenocorticotrophin) and adrenal (hydrocortisone) glands control their activities during early development.

Mechanism

Although the mechanism of disaccharide membrane digestion is not yet clear, several attractive hypotheses have been postulated to explain the process.

In 1959 A. M. Ugolev suggested that pancreatic α-amylase in the intestinal lumen became adsorbed onto the absorptive cell membrane and luminal starch was then digested on the outer cell surface *(contact digestion)* before being absorbed. Ugolev later (1965) proposed that digestion of carbohydrate molecules was initiated by pancreatic α-amylase trapped in the glycocalyx and then completed by disaccharidases adsorbed onto the external surface of the microvilli. Initially, the disaccharidases were synthesized intracellularly and subsequently released from the cells; hence the digestive process is extracellular. In 1968 J. D. Hamilton and H. B. McMichael suggested that the glycocalyx provided a mechanism for the efficient trapping of sugar molecules which became digested by disaccharidases located within the surface coat. Mobile carriers then transported the monosaccharide-sodium linked molecules across the microvillous membrane into the cell.

In 1962 R. K. Crane showed that the hydrolysis of disaccharides occurred within the membrane at a site superficial to the transport mechanism because phloridzin in the lumen did not prevent the hydrolysis of sucrose but stopped the absorption of released glucose. The monosaccharides that are formed by hydrolytic action are then transported by mobile carriers, coupled with sodium molecules, at a site in the membrane internal to that of disaccharidase activity, to be released inside the cell. Figure 12.4 *(left)* summarizes the mode of action of the hydrolysis of a molecule of sucrose and the subsequent transport of the glucose, fructose and sodium molecules through the absorptive cell.

In 1967 C. F. Johnson isolated several disaccharidases from 'knobs' (p. 71) covering the microvillous membrane, and a year later T. Oda and S. Seki identified membrane ATPase in the inner protein leaflet of the membrane. Thus, it is possible that disaccharides are first hydrolyzed in the 'knobs' and the products of hydrolysis are then transported by carriers, the action requiring catalysis by membrane ATPase deeper in the membrane—an elegant correlation between ultrastructure and biochemical function at the molecular level. Figure 12.4 *(right)* illustrates this mechanism. However, the 'knobs' seen under the electron microscope following hydration of the tissue are possibly artefacts. Moreover, disaccharidases have been recently isolated from the microvillous membrane after removing the 'knobs' by papain (enzyme) digestion. Clearly, the mechanism for disaccharide membrane digestion remains as yet unresolved.

Sugar (monosaccharide) absorption

Before being absorbed into the bloodstream and transported to the liver disaccharides are all hydrolyzed to monosaccharides on the surface or, more probably, within the microvillous membrane, yielding 80 per cent glucose, 15 per cent fructose and five per cent galactose, or about 2000 millimoles of monosaccharides are released.

Sugars are mainly absorbed in the small intestine; sugar absorption in the stomach and large intestine is very limited. Many sugars enter the mesenteric blood at a concentration lower than in the intestinal lumen, that is, they are absorbed along a concentration gradient. Thus, after eating a typical meal, about 800 mg of glucose per 100 cm³ fluid can be recovered in the jejunum, well above the 100 mg of glucose per 100 cm³ blood levels in the plasma. Although sugars can pass with equal ease either into the blood capillaries or lymph channels they are swept into the bloodstream since the flow there is far greater than in the lymphatics.

Depending on the species, the *site* for active monosaccharide transport is either mainly in the jejunum or in the ileum (Fig. 12.5), and the absorptive capacity is possibly related to the site of maximum disaccharidase activity. Some sugars hydrolyzed in the microvillous membrane diffuse back into the lumen to be absorbed further down the small intestine. The *rates* at which different monosaccharides are absorbed differ. Thus, equal concentrations of galactose, glucose, fructose, mannose, xylose or arabinose are absorbed by the rat small intestine in decreasing order of magnitude. The rate of sugar absorption depends on

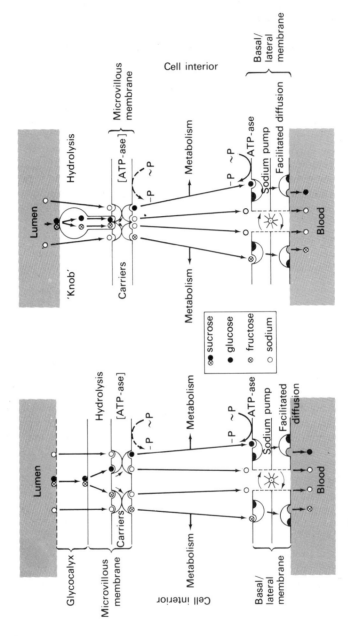

Fig. 12.4   Functional relationships in the absorptive cell. Details in text.

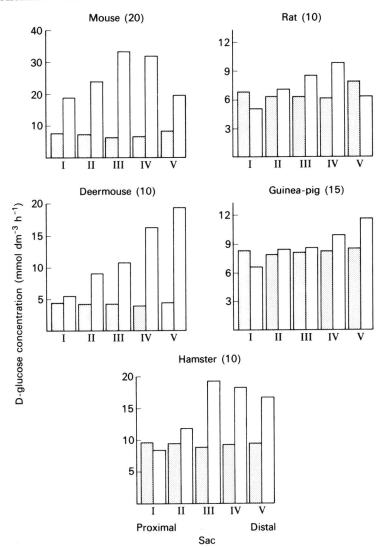

Fig. 12.5 Absorption of D-glucose from different segments of equal length of the entire small intestine in laboratory mammals using *in vitro* everted sacs. Sac 1: proximal end; sac V: distal end. Each paired histogram gives on thè *left* the final mucosal concentration and on the *right* the final serosal concentration after incubation for 1h at 37°C. Initial mucosal or serosal concentration: 10.0 mmol dm^{-3}. The low, final serosal glucose concentration in the rat and guinea-pig results from the relatively high rate of glucose metabolism by the tissues during transport. Number of animals used in brackets (adapted from Madge, 1972).

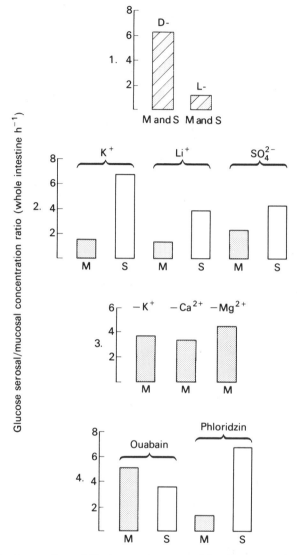

Fig. 12.6 Aɒsorption of D-glucose by everted *in vitro* sacs of entire small intestine of the mouse after 1 h at 37°C under different experimental conditions. A final serosal/mucosal concentration ratio of 1.0 or greater indicates that absorption occurred. (1) The D-isomer but not the L-isomer is absorbed. (2) When sodium ions are *replaced* by either potassium or lithium ions in the mucosal (M), but not the serosal (S) fluid, absorption is virtually abolished. When chloride ions are replaced by sulphate ions absorption is also lowered. (3) When potassium, calcium or

their molecular configuration and not on their size or solubility (p. 112).

The mechanism for intestinal absorption of many sugars at low concentrations involves *active transport* of the molecules and hence maintains an electrochemical gradient, for in the presence of certain metabolic inhibitors the absorption of glucose or galactose, for example, but not that of ribose or arabinose, is abolished. However, limited passive transport of actively absorbed sugars in the jejunum is probable. Lack of oxygen or low temperature also stops absorption. The importance of *sodium ions* for active *in vitro* absorption at low sugar concentrations has already been stressed (Chapter 10). At high sugar concentrations, however, the same maximal rate of sugar transfer is attained irrespective of the sodium concentration. Under *in vivo* conditions in man sodium ions do not appear to be essential for the absorption of glucose. However, sodium secreted by the absorptive cells may accumulate in the glycocalyx and 'unstirred' water layer, thus providing an unsuspected pool of these ions for the transport process (p. 89). When sodium ions are replaced by potassium, lithium, magnesium, mannitol or Tris, active *in vitro* absorption of sugars is either greatly lowered or abolished. Also, if certain ions (potassium, chloride, magnesium, calcium) are omitted from normal saline-bicarbonate buffer in the mucosal fluid active absorption becomes depressed. Figure 12.6 gives the results of experiments summarizing some of the important requisites for active absorption of glucose by the mammalian (mouse) small intestine.

Different sugars compete with each other for the carrier site. If present together in the lumen they often mutually *inhibit* one another (competitive inhibition). Thus, if glucose and galactose are present initially in equal concentrations in the rate intestine, there is a reduction in glucose absorption of about 15 per cent and in galactose absorption of about 50 per cent. The fact that monosaccharides can partially inhibit one another indicates that there is a common pathway on the mobile carriers shared by them. The presence of amino acids can also interfere with or affect the capacity of monosaccharide transport, suggesting that such a carrier can also be shared by amino acids. For this reason, the carrier is sometimes referred to as a *polyfunctional carrier*. Alternatively, the limited endogenous supply of the cell may be a limiting factor (p. 123). Although a polyfunctional carrier mechanism is possible it is difficult to reconcile the observations that, in some mammals, monosaccharides are absorbed mainly by the

---

magnesium ions are *omitted* from normal Ringer bicarbonate-saline solution in the mucosal fluid (M) absorption of glucose is decreased. (4) Ouabain (5×10^{-5} mol dm^{-3}) in the serosal (S) but not the mucosal (M) fluid decreases transport (this change is much less apparent in the mouse than in some other animals); phloridzin (2×10^{-4} mol dm^{-3}) in the mucosal (M) but not the serosal (S) fluid stops absorption. Mean results of six animals for each experiment. Initial glucose concentration in both mucosal and serosal fluids: 10.0 mmol dm^{-3}. (Origin).

jejunum whereas amino acids are mainly absorbed by the proximal ileum, and hence at different sites along the intestine. Current work indicates that more than one glucose *carrier-type* with selective action may operate in the rat jejunum and ileum. Finally, the rate at which sugars are actively absorbed is not directly proportional to their concentration in the intestinal lumen, as it is with sugars that are absorbed by simple diffusion, but approaches a limiting value as the concentration is increased. The active absorption process is said to exhibit *saturation* (Michaelis–Menten) *kinetics,* like enzyme catalyzed reactions (Fig. 10.2).

Fructose is found in fruits and is part of the sucrose molecule. During absorption, the metabolic pathway followed by fructose depends on the animal studied (Fig. 12.7). In man, fructose recovery is virtually complete. In the rat 20 per cent of the original fructose is recovered as lactic acid, ten per cent is converted to glucose and the remainder is unaltered; in the mouse more fructose is converted to lactic acid. Hamster and guinea-pig small intestine have both fructokinase and glucose-6-phosphatase in the mucosa which convert over 50 per cent of the fructose into glucose. In hamster and mouse everted intestinal sacs fructose recovery in the serosal fluid after incubation is significantly lower than that in the mucosal fluid owing to the extensive metabolism of fructose into glucose and lactic acid during transfer.

Until recently it was accepted that fructose was absorbed by either simple or facilitated diffusion. However, in 1970–72 M. Gracey and colleagues showed that D-fructose at low concentrations is actively transported. Fed rats absorbed relatively small amounts of fructose but starved animals absorbed more. The marked inhibitory effects of metabolic poisons, lack of oxygen or low temperature all indicate that the transport mechanism depends on metabolic energy, and hence transport is an active process. Fructose absorption is sodium-dependent for replacing sodium with potassium or Tris significantly reduces transport. (Other workers have found, however, that replacing sodium with potassium will result in moderate fructose absorption). Fructose absorption also exhibits saturation kinetics and some competitive inhibition with other sugars and amino acids.

Since fructose absorption is not inhibited by phloridzin the carrier mechanism involving fructose transport probably differs from that of glucose, although the process is probably similar. Also, the molecular configuration of fructose is unlike the minimal structural requirements for active sugar transport (below).

Specificity for active sugar transport

In 1960 R. K. Crane first defined the minimal structural requirements for active sugar transport. These requirements were a pyranose (six carbon) ring, a free hydroxyl group attached at C-2 in the glucose configuration, and an additional carbon (C-6) fixed to C-5 (a methyl or methyl substituted group). Later, the D-isomer was also considered essential for active transport.

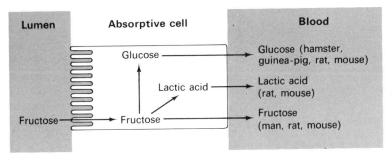

Fig. 12.7 Fructose absorption and conversion in the absorptive cell in different mammals (adapted from Wilson, 1962).

However, it was subsequently shown that the carbon attached to C-5 was not essential; D-xylose, for example, shows limited active absorption. Moreover, the L-isomer of glucose is also slightly absorbed when rats and hamsters, but not mice, were fed on a restricted diet. (L-isomers of sugars, like D-isomers of amino acids, are neither produced nor utilized by animal cells.) The hydroxyl group at C-4 in L-glucose has a similar configuration to that at C-2 in D-glucose. If this specificity is expressed by numbering the pyranose ring from the oxygen atom, then the hydroxyl group at position 3 includes both C-2 in D-glucose and C-4 in L-glucose (Fig. 12.8). This is the minimal entry specificity for sugars, although an additional carbon at C-5 greatly increases absorption. Recent evidence suggests that hydrogen bonding at C-1 and C-6 is also necessary for active intestinal sugar transport.

D-fructose, a sugar with a different molecular configuration to other actively absorbed sugars, is also actively absorbed. Thus, the carrier involved in fructose transfer probably differs from that of the other actively transferred sugars.

Energy supply for sugar transport

When studying sugar absorption, D. H. Smyth in 1971 stressed that a distinction must be made between sugars that *utilize* energy and sugars that *supply* energy. Glucose is both transferred and metabolized by the cell while galactose is transferred by not significantly metabolized. Therefore, when studying the intestinal transfer process investigations using galactose, an energy user, is more useful than using glucose, which is both an energy user and supplier.

Galactose in the mucosal fluid is actively absorbed, but in the additional presence of glucose the absorption of galactose is lowered since one sugar competes with the other for the carrier site. However, if glucose is put into the serosal fluid it enters the absorptive cells through the serosal surface without competing for entry with galactose, providing additional energy for the cells. Under these

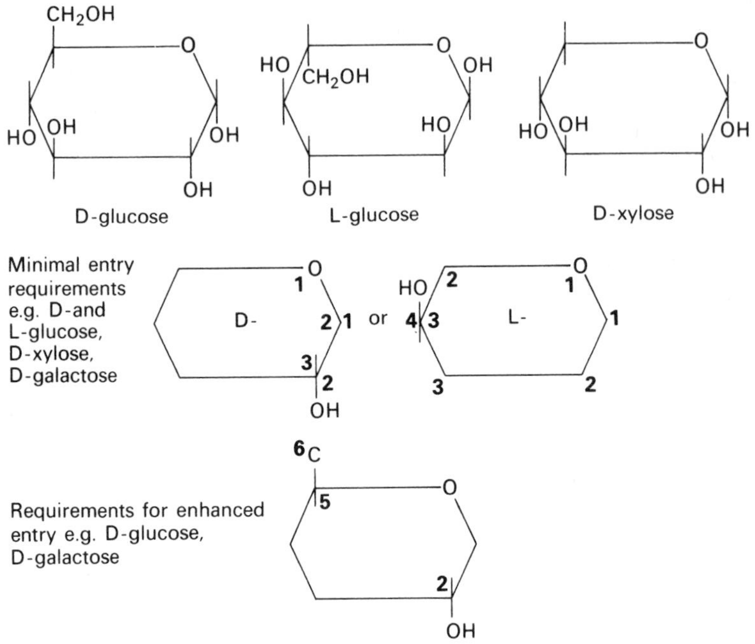

Fig. 12.8 Structural entry requirements for actively absorbed sugars. Further details in text (after Smyth, 1971).

conditions absorption of galactose is appreciably enhanced, although not at a maximal rate. Therefore, galactose transport can be energized to some extent by endogenous (cell) metabolism, but the exogenous (blood or serosal) energy supply is more important. Incidentally, glucose in the serosal fluid will also enhance amino acid uptake from the mucosal fluid by energizing amino acid entry.

The *biochemical pathways* for hexose metabolism have been studied by using selective metabolic inhibitors in the mucosal fluid. Fluoride affects the glycolytic cycle and fluoroacetate affects the tricarboxylic acid cycle. In the absence of glucose in the serosal fluid fluoroacetate, but not fluoride, in the mucosal fluid in-hibits the transfer of galactose. Therefore, the endogenous energy for galactose transfer must normally mainly come from the tricarboxylic acid cycle. Since this energy cannot come from the metabolism of stored carbohydrates (the starting point for carbohydrate metabolism is glucose) it must come mainly from a non-carbohydrate source, for example, fat metabolism.

The increased transfer of galactose following stimulation by glucose in the serosal fluid is inhibited by fluoride, but not by fluoroacetate, in the mucosal fluid. Hence, the energy for the increased galactose uptake must mainly come

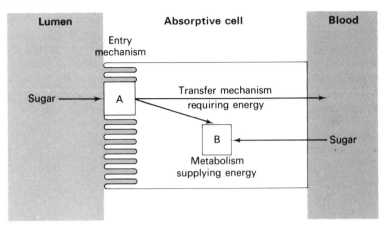

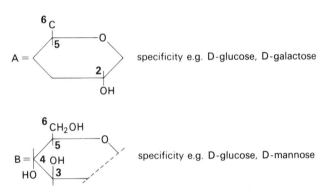

Fig. 12.9 Specificities for sugar entry (A) and sugar metabolism (B) in the absorptive cell (adapted from Smyth, 1971).

from the glycolytic cycle and not from the tricarboxylic acid cycle. So, although the tricarboxylic acid cycle and the glycolytic cycle are both used during galactose transfer, the glycolytic pathway appears to be more important when glucose is present (in serosal fluid or blood).

Specificity for sugar metabolism

The minimal structural requirements for the metabolism of hexoses include carbon atoms attached to positions 3, 4, 5 and 6 and the oxygen atom of the pyranose ring (Smyth, 1971). This has been called the *hexokinase specificity,* as distinct from the *entry specificity* (or Crane Specificity) (Fig. 12.9), and includes, as examples, D-glucose and D-mannose. These sugars are metabolized and are therefore energy suppliers. The entry specificity, as already indicated, consists of a pyranose ring with a hydroxyl group at C_2 and a methyl group attached to C_5, and includes, as examples, D-glucose and D-galactose. Figure 12.9 represents the location of the specificities related to sugar entry and metabolism in the absorptive cell.

13   Amino acids and proteins

Amino acids

Proteins are large molecules formed from amino acids linked together by peptide (—NH—CO—) bonds. These bonds are formed by joining the acidic or amino (—NH$_2$) group of one amino acid to the basic or carboxyl (—COOH) group of another amino acid. Proteins are the most important of all dietary constituents since they are the basic components of protoplasm in the cell. Also, all enzymes and many hormones are proteins. To maintain a stable nitrogen balance, an adult human needs about half a g of high quality protein for each kg of body weight per day. Endogenous proteins in the form of digestive secretions and shed epithelial alimentary canal cells may provide up to half the total protein requirements.

Extracellular digestion

Dietary proteins are partially digested in the stomach by pepsin and in the duodenum by trypsin, chymotrypsin and elastase found in the pancreatic juice, splitting inner peptide bonds into a complex mixture of large peptide fragments. Pancreatic juice also contains carboxypeptidases which break off terminal carboxyl bonds, and some aminopeptidases which fragment terminal amino links. Proteins may become hydrolyzed by protein enzymes trapped within the glycocalyx and overlying 'unstirred' water layer. Pancreatic proteases generally yield a low concentration (about one millimole per litre) of 10–25 per cent free neutral and basic amino acids and a mixture of 75–90 per cent small peptides, or *oligopeptides,* of different (2–6) chain lengths. Free amino acids in the intestinal lumen are absorbed by the epithelial cells but oligopeptides enter the cells and become hydrolyzed mainly by specific di-and tripeptidases.

Membrane and intracellular digestion

Although there are about 400 possible dipeptides and 8000 possible tripeptides, very few peptide hydrolases have yet been isolated, and hence little is known about cellular digestion of oligopeptides. H. Newey and D. H. Smyth in 1957 first reported the uptake of intact oligopeptides by the absorptive cells and their subsequent hydrolysis within the cells. From 1968 D. M. Matthews and colleagues showed that the uptake of some tripeptides is more rapid than the uptake of some dipeptides, and that of some dipeptides more rapid than that of equivalent free amino acids (Fig. 13.1). Moreover, when the amino acids were presented as a mixture, there was greater competitive inhibition of the free amino acids than when presented as the corresponding di- or tripeptide. These and other observations suggest that some oligopeptides are taken up intact by specific peptide transport sites on the microvillous membrane of the epithelial cells, that is, transport within the cells precedes their hydrolysis.

Since some oligopeptides (for example, glycylglycine) have molecular weights smaller than most free amino acids it is perhaps not surprising than such oligopeptides are capable of being transported across the cell barrier. It is possible, however, that oligopeptides, like disaccharides (p. 106), are also hydrolyzed superficially in the glycocalyx of the microvillous border (contact digestion), with

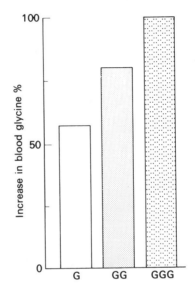

Fig. 13.1   Increases in blood glycine after equivalent oral doses of glycine (G), glycylglycine (GG) or glycylglycylglycine (GGG) in man (modified from Matthews, 1972).

simultaneous attachment of amino acids to transport sites. *In vitro* experiments have shown that some oligopeptides require sodium for their entry into the cell and compete with each other for their entry route. They show evidence of saturation kinetics and transport is reduced by oxygen deprivation and metabolic inhibitors, suggesting that an active, sodium-dependent, energy-requiring mechanism is required for their transfer. Oligopeptide absorption also exhibits stereospecificity for L- rather than D-isomers are absorbed.

The *development* of peptidases in the growing animal has received little attention. In pre-natal rats peptidase activities are very low but rapidly increase for a short time in new-born animals when their activities become about three times greater than in adults. The reason for this transient rise in peptidase activities at birth is unknown, but it has been suggested that inhibitory factors in the colostrum retard their initial activities and may prevent their subsequent development. In man, by contrast, the levels of both the pre- and post-natal small intestinal peptidases remain unchanged. Isolation of peptidases from different regions of the small intestine in adults suggests that generally the main *site* of their activities is in the proximal region. The stomach and large intestinal epithelium have low peptidase activities, as do the secretions in the alimentary tract lumen.

Using sub-cellular fractionation techniques T. J. Peters (1970) isolated tripeptide hydrolases mainly in the plasma membrane of the microvilli and dipeptide hydrolases mainly in the cell cytoplasm. Of ten dipeptidases identified, only five to ten per cent of their total activities were isolated in the microvillous membrane and most of the remaining activities were found in the cytoplasm. Of five tripeptidases recovered, up to two-thirds of their activities were confined to the apical membrane and about one-third were found in the cytoplasm.

Figure 13.2 summarizes the speculative mode of protein digestion in adult mammals. Oligopeptides from both exogenous and endogenous proteins are released in the intestinal lumen by gastric and pancreatic proteases. Part of this digestion is probably also brought about by luminal enzymes adsorbed onto the surface of the microvillous membrane. Free amino acids are transferred mostly unchanged in the blood, although a few (e.g. arginine, asparagine, ornithine) are metabolized during transfer. Tripeptides are hydrolyzed to dipeptides and free amino acids by apical membrane tripeptidases. Amino acids and dipeptides are then transported into the cell, where intracellular dipeptidase hydrolysis occurs. Finally, free amino acids cross the basal membrane and pass into the portal circulation. A small proportion of oligopeptides may enter the bloodstream unhydrolyzed. However, the fact that several dipeptidases and tripeptidases have overlapping substrate specificities complicates this scheme. Details of the mechanism of oligopeptide absorption are still very incomplete and much more work remains to be done before definitive transport models can be made.

Amino acid absorption

There are over twenty different amino acids, grouped as neutral, acidic or basic (Table 13.1). About half of them cannot be synthesized by the animal and are therefore indispensable in the diet; they are called 'essential' amino acids. Amino acids do not usually exist in the diet in free form, but a small proportion of proteins digested by pancreatic proteases in the intestinal lumen may enter the absorptive cells as free amino acids. The *concentration* of amino acids in the intestinal lumen is affected by the rate of release of free amino acids by proteases, to-and-fro movements of molecules by exchange diffusion (below), the back-

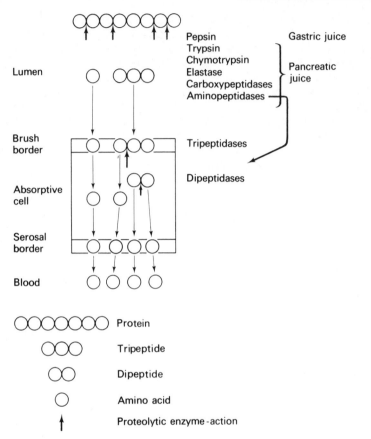

Fig. 13.2   Digestion of proteins in the small intestine of an adult mammal (tentative).

diffusion of amino acids liberated by intracellular peptidases, and by the widely differing rates of absorption of the free amino acids.

Many amino acids are *actively absorbed* by a carrier-mediated process and are thus capable of maintaining an electrochemical gradient across the cell, moving amino acid molecules 'uphill' against a cell to lumen concentration gradient (Fig. 10.6). Since amino acids accumulate inside the cell while transport proceeds, it is possible that an active transport 'pump' mechanism is situated in the apical membrane. With few exceptions, only the L-forms are generally actively transported, which are the naturally occurring isomers in the diet. The sodium molecule is essential for the absorption of many amino acids but not for some as,

Table 13.1   Classification of amino acids based on the nature of their side chains. Essential amino acids are italicized. (The imino group are not true amino acids since each amino group is not free).

1 NEUTRAL
  (a) Aliphatic group
     Alanine, glycine, *isoleucine*, *leucine*, serine, *threonine*, *valine*

  (b) Aromatic group
     Phenylalanine, *tryptophane*, tyrosine

  (c) Sulphur-containing group
     Cysteine, cystine, *methionine*

  (d) Imino group
     Betaine,* hydroxyproline, proline, sarcosine*

2 ACIDIC
  Aspartic acid, glutamic acid

3 BASIC
  Arginine, citrulline, *histidine*, hydroxylysine, *lysine*, ornithine

* Not true constituents of proteins.

for example, glycine, lysine or proline. Vitamin B_6 (pyridoxine) may also enhance normal amino acid transport across the cell. The energy for transport involves oxidative metabolism; certain poisons, for example dinitrophenol or cyanide, and oxygen deprivation or low temperature will all inhibit absorption. Since ouabain in the serosal fluid also usually stops absorption, membrane ATPase appears necessary for sodium transfer. The exit of amino acids from the basal membrane is probably by non-sodium dependent facilitated diffusion. Figure 13.3 summarizes some of the conditions necessary for active transport of amino acids in the mammalian (mouse) small intestine.

Several *transport systems* have been described for amino acid transfer. Neutral amino acids require two different carriers, the so-called methionine and sarcosine carriers. The methionine carrier transports methionine, alanine, leucine, glycine, proline and valine, prefers long-chain amino acids, and is specific for α-amino acids and L-isomers. The sarcosine carrier cannot transport methionine or leucine but transports sarcosine and betaine (both imino groups), glycine and proline. This carrier prefers short-chain amino acids, handles amino acids which are not in the α-position, and transports either L- or D-isomers. Other less well-defined carrier systems have been described for acidic and basic amino acids.

A number of amino acids will *compete* with each other for the carrier site, some being better inhibitors than others. For example, in rats the transport of L-histidine is inhibited by about 90 per cent, 60 per cent or 20 per cent in the presence of L-leucine, L-valine or glycine respectively. The presence of certain monosaccharides will also decrease amino acid transport, suggesting a multifunctional property of the carrier (p. 111). However, monosaccharides are more

122

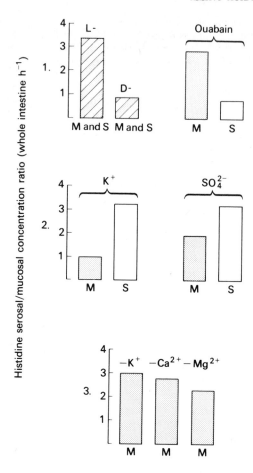

Fig. 13.3 Absorption of L-histidine by everted *in vitro* small intestinal sacs of the mouse after 1 h at 37°C under different experimental conditions. A final serosal/mucosal concentration ratio of 1.0 or greater indicates that absorption occurred. (1) The L-isomer but not the D-isomer is absorbed. Ouabain (5×10^{-5} mol dm^{-3}) in the serosal (S) but not the mucosal (M) fluid virtually stops absorption. (2) When sodium ions are *replaced* by potassium ions in the mucosal (M) but not in the serosal (S) fluid absorption is stopped. When chloride ions are replaced by sulphate ions absorption is decreased. (3) When potassium, calcium or magnesium ions are *omitted* from normal Ringer bicarbonate-saline solution in the mucosal fluid, absorption of histidine is slightly decreased. Mean results of six animals for each experiment. Initial histidine concentration in both mucosal and serosal fluids: 2.0 mmol dm^{-3}. (Original).

likely to inhibit amino acid transport by competing for the limited energy available in *in vitro* preparations (Fig. 10.2). For example, with mannitol (a non-metabolizable substance) in the serosal fluid, galactose partially inhibits the transfer of glycine, suggesting competitive inhibition at the carrier site. But with mannose (a metabolizable sugar) in the serosal fluid (p. 96), the effect of galactose on glycine absorption is abolished, showing that in fact galactose acts as a non-competitive inhibitor, competing for the available energy.

A few amino acids, like L-aspargine, L-leucine and L-tyrosine, show in addition limited *exchange diffusion,* when a small proportion of analogous amino acid molecules move back and forth between cell and lumen. However, the net influx of molecules will eventually exceed the net efflux, resulting in net absorption. Finally, with at least two exceptions, D-methionine and D-alanine, which are actively transported, D-amino acid isomers move through water-filled channels by *simple diffusion.*

Like monosaccharides, the *absorption rate* of different amino acids varies. Thus, the rate of transport of several amino acids in the human jejunum decreases in the following order: L-valine, L-alanine, glycine, L-leucine and L-methionine. There is a route for amino acid entry along the small intestine which is distinct from the route of oligopeptide entry (Fig. 13.4). The absorption of some amino acids in certain species is mainly limited to the proximal ileum but that of oligopeptides is mainly in the jejunum, suggesting that the uptake of amino acids and that of oligopeptides are independent of each other and so are transported by different carriers. Moreover, M. D. Milne and his associates

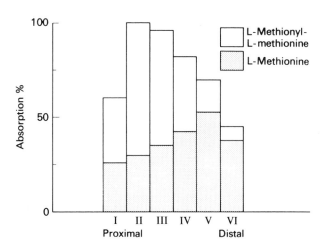

Fig. 13.4   Absorption of L-methionine and L-methionyl-L-methionine along rat small intestine (adapted from Matthews, 1972).

(1972) have found that in rare inherited disorders of metabolism, such as Hartnup disease or cystinuria, active transfer of free amino acids, particularly neutral amino acids, is virtually abolished although the corresponding oligopeptides are transferred normally, providing further evidence that oligopeptide uptake is a separate process to that of free amino acids. Thus, in some mammals any amino acid molecules liberated by protein digestion in the lumen of the proximal small intestine will be eventually absorbed in the distal region.

Specificity for active amino acid transport

The basic pattern of an amino acid molecule is

$$R-\overset{\displaystyle \overset{H}{|}}{\underset{\displaystyle \underset{NH_2}{|}}{C_a}}-COOH$$

R is a hydrogen atom
or a complex radical

(1) Except for glycine and β-analine, amino acids exist as either L- or D-isomers. Only the L-isomer is usually actively absorbed in appreciable amounts. Neutral amino acids are easily absorbed against a concentration gradient, even at high concentrations. Basic amino acids are, however, easily absorbed only at low concentrations. Acidic amino acid absorption is difficult to demonstrate since they are rapidly deaminated during transfer. Only a few D-isomers (e.g., D-methionine and D-alanine) are actively absorbed and most other D-forms passively absorbed. Since the actively absorbed forms of the neutral L-amino acids reduce the passive transport of their D-forms, it can be inferred that both isomers share a common mobile carrier. (2) If the carboxyl group in the molecule is substituted, active absorption ceases. Therefore, the intact —COOH attached to the a-C atom is essential for transport. (3) The a-amino group may not be essential, for if the a-NH$_2$ radical is substituted by a hydrogen atom there is very limited absorption. (4) The a-H atom is also not essential, for when replaced by an a-CH$_3$ group there is some absorption. (5) When the side chain R- is substituted absorption is markedly reduced as, for example, in L-lysine or L-arginine; if missing as, for example, in L-aspartic acid, absorption virtually stops.

Proteins

Unlike the adult the young mammal is able to absorb proteins which confer upon it *passive immunity,* giving it transient protection against invading pathogens (organisms that cause infection) by producing *antibodies* which are induced by alien molecules called *antigens (immunogens)* forming part of, or produced by, the pathogens. Large antigen molecules pass into the villi mainly through the desquamated tips, while smaller ones may enter the absorptive cells by an active transport mechanism. Initially, the young acquire antibodies from the mother in the form of a group of large, intact, immune proteins called *immunoglobulins,* either before birth in the bloodstream across the yolk sac or placenta, or by intestinal absorption of immunoglobulins in the colostrum (initial secretions from the mammary glands) soon after birth. Man, rabbits and guinea-pigs acquire immunoglobulins before birth; domestic animals and rodents acquire them both before and (mainly) after birth, and farm animals acquire them shortly after birth. Immunoglobulins in the alimentary canal escape hydrolysis either because proteolytic enzyme levels are low in the new-born or because the colostrum in certain species contains enzyme inhibitors. Soon, however, the mechanism for protein absorption stops and the young then produces its own antibodies in response to antigens, conferring *active immunity* on the growing animal.

Immunoglobulins form part of blood serum which consists of four main components: albumin, and alpha-, beta- and gamma-globulins. Gammaglobulins having antibody activity are called immunoglobulins. In man, there are five classes of immunoglobulins (Ig): A, D, E, G and M. An immunoglobulin molecule is a glycoprotein consisting of a carbohydrate moity and four long polypeptide chains, two 'light' ones (M.W. 22 000) common to all five classes and two 'heavy' ones (M.W. 65 000) specific to each class, linked to each other by di-sulphide bridges. The heavy chains contain antibody sites while the light chains are relatively inactive. After acquiring the antigen specialized plasma cells (*plasmocytes*) and large lymphocytes (*immunocytes*), originating mainly from the spleen and lymph nodes (Peyer's patches), synthesize and release into the blood a specific antibody (immunoglobulin) to the particular antigen. Any one individual blood cell makes only one class of immunoglobulin.

Immunoglobulin M is the first one to appear in response to an invading pathogen. IgM is present in only small amounts and soon disappears from the circulation, to be replaced by IgG, the most abundant immunoglobulin. The IgA molecules probably appear later than IgM but earlier than IgG. IgA is also synthesized in 'local' plasma cells of the lamina propria, then escapes into the subepithelial spaces in the villi, to either enter the absorptive cells, possibly by pinocytosis or pass into the intercellular spaces. Two molecules of IgA combine with a 'secretory piece' or 'transport piece' glycoprotein, to be secreted intact by the absorptive cell into the overlying mucus. The mucous coat thus eventually acquires secretory (local) IgA antibodies distinct from serum (systemic) IgA antibodies, conferring upon it mild antiseptic properties and so deterring viruses, bacteria and helminth parasites from invading the intestinal epithelium (IgA is also found in the mucus lining the respiratory tract, in saliva and in tears). The mode of action of IgA is obscure. IgE is concerned with allergic skin and respiratory reactions and is involved in the reaction to helminths. IgM is effective against Gram-negative bacteria. IgD has no known function.

In the new-born colostral proteins including immunoglobulins are readily taken up by the absorptive cells by pinocytosis in the mid- or distal small intestine. Although the main colostral immunoglobulin is usually A, only IgG is easily transported intact, while colostral IgA, like secretory IgA later, becomes adsorbed onto the villous surface. IgG complexes of high molecular weight are first formed which, unlike other proteins, become attached to specific protein receptors on the plasma membrane at the base of the microvilli. The IgG molecules are carried into the cell by the incorporation of the membrane into pinocytotic vesicles which communicate with apical tubules, forming a tubular-vesicular system. The vesicles then break away from the microvillous membrane and membrane-bound IgG is released inside the cell. Serum albumin is also taken pinocytotically, although less readily, than immunoglobulins. The subsequent route of the vesicles depends on the species studied.

In ruminants, pigs and carnivores within one hour of birth the apical cytoplasm of the absorptive cell becomes filled with small microglobules of protein within membrane-covered vesicles (Fig. 13.5). The vesicles move towards the base of the cell, fusing as they do so, by-passing the nucleus and displacing it towards the cell apex. Eventually a large, membrane-bound, protein-filled macroglobule, about 10 μm in diameter, is formed at the base of the cell (Fig. 13.6a). After six to nine hours the globule escapes into the lymphatic system, possibly by 'reverse' pinocytosis at the lateral plasma membrane and through the dilated intercellular spaces. The cells thus absorb a maximum amount of protein before expelling it at a constant rate into the circulation. The transfer of proteins across the absorptive cells is sometimes facilitated by the presence of different specific substances in the colostrum. Unlike other mammals young ruminants, pigs and carnivores take in proteins non-selectively, that is, both immunoglobulins and albumin are absorbed.

In young rodents the initial protein uptake is similar to that in domestic animals. But after two hours a large, complex, diffuse vacuolar globule surrounded by a smooth membrane is formed above the nucleus. This vesicle is probably formed by the fusion of pinocytotic vesicles and lysosomes. The contents of the supranuclear body thus include both exogenous and endogenous materials. The exogenous fraction consists of immunoglobulin, and the endogenous fraction includes lysosomal acid hydrolytic enzymes and acid phosphatase (Fig. 13.6b). The supranuclear body does not change its position and its contents are eventually discharged into the intercellular spaces. In rodents, proteins are taken in selectively, that is, only immunoglobulins are absorbed.

In the human foetus 'meconium corpuscles' derived from proteins in the amniotic fluid are seen in the apex of the absorptive cells in the ileum. Although these bodies appear to be analogous to the supranuclear vacuoles found in young rodents, they are apparently not released into the circulation and hence their functions are obscure. Passive immunity (IgG antibodies) from mother to foetus normally occurs across the placental epithelial cells by pinocytotic vesicles released from microvilli.

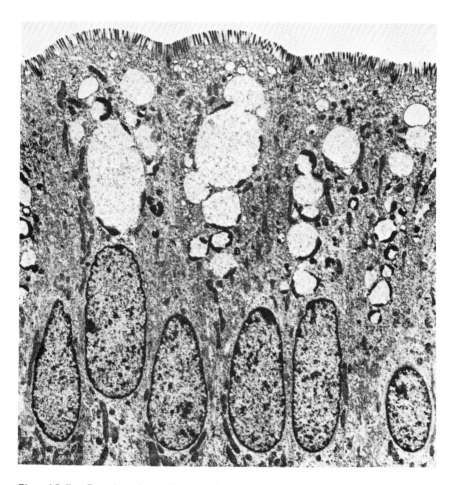

Fig. 13.5 Protein absorption in ileal absorptive cells of the young pig. Supranuclear vacuoles in the apical cytoplasm contain granular material and peripheral clumps of electron-dense material. Transmission electron microscopy. (Reproduced by permission from Moon, H. W. (1972) *Vet. Path.* **9,** 11.)

Recent work suggests that the absorption of unhydrolyzed immunoglobulins by pinocytosis in the young is more complex and specific than at first thought. Their transfer is associated with sodium and water uptake, shows competitive inhibition with amino acids, is affected by metabolic antagonists and requires oxygen. Thus, some form of an active transfer mechanism is probably involved in their uptake.

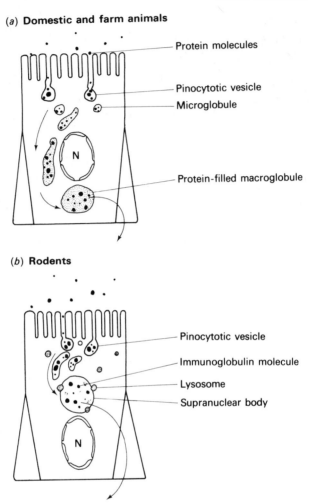

(a) Domestic and farm animals

— Protein molecules

— Pinocytotic vesicle
— Microglobule

— Protein-filled macroglobule

(b) Rodents

— Pinocytotic vesicle

— Immunoglobulin molecule

— Lysosome
— Supranuclear body

Fig. 13.6   Protein uptake by the absorptive cell in the young mammal: (a) domestic and farm animals, (b) rodents.

The formation of pinocytotic vesicles involving protein absorption irreversibly stops within a few days when the immunoglobulin concentration in the young equals the adult level. In ruminants and guinea-pigs vesicle formation ends after one to one and a half days, in hamsters six days, in dogs about twelve days, in rodents about eighteen days and in rabbits twenty three days. There seems to be no relationship between the turnover times of the villous epithelium and the end of vesicle formation. The factors which end the transmission of passive immunity after birth are complex and little understood, although possibly related to a solid diet, hormonal stimuli and the onset of gastric secretion. There is some evidence to show that the mature intestinal cells retain a limited capacity to absorb intact proteins, but protein absorption in adults frequently results in allergic reactions to certain foods.

14   Fats, cholesterol and bile salts

Fats

The dietary intake of fats in humans is about 70 g each day; endogenous fats from the desquamated small intestinal epithelium provide up to one-fifth of that amount. Dietary fats are important not only for their high energy value but also because they insulate subcutaneous tissues and various organs. Ingested fats include mainly neutral fats (triglycerides) and small amounts of phospholipids, steroids and free fatty acids. A triglyceride molecule consists of three fatty acids bound to glycerol. Steroids include steroid hormones and sterols, mainly cholesterol. Fatty acids are either saturated or unsaturated (dehydrogenated) and obtained from the hydrolysis of fats. Other compounds associated with fats include oils, waxes, bile acids and fat-soluble vitamins. The digestion and absorption of fats can be divided into several phases.

Extracellular phase

In the monogastric animal extracellular digestion is achieved by one enzyme-lipase. The stomach secretes a weak lipase, and less than ten per cent of total dietary fats (mainly as medium-chain fatty acids) are digested there. Fats are mainly digested by calcium-activated pancreatic lipase in the proximal small intestine, which first hydrolyzes triglycerides as long-chain fatty acids to diglycerides, then diglycerides to monoglycerides, free fatty acids and glycerol (Fig. 14.1). Medium-chain fatty acids are completely hydrolyzed to free fatty acids and glycerol. Although not essential for triglyceride absorption, bile salts increase the speed and efficiency of absorption of the hydrolytic products. Thus, bile salts solubilize fats in the intestinal lumen, emulsify triglycerides and diglycerides, help in the formation of micelles, and displace the pH optimum of lipase from 8 to 6. In the small intestinal lumen monoglycerides, fatty acids, glycerol and bile salts combine to form unstable, negatively charged polymolecular aggregates, *micelles,* each about 50 Å in diameter.

In the adult digastric animal fats are digested in the rumen. The food eaten contains triglycerides, consisting mainly of long-chain fatty acids, three quarters of which are unsaturated fatty acids (mainly linolenic acid). Bacteria and protozoa in the rumen (p. 40) ferment the unsaturated fatty acids by hydrogenation and hydrolysis to milk fats rich in saturated fatty acids (mainly stearic and palmitic acids). Triglycerides are also hydrolyzed to free fatty acids and glycerol; unlike non-ruminants monoglycerides and diglycerides are not produced. The adult ruminant, however, can metabolize unsaturated fatty acids when introduced directly into the proximal small intestine, thereby by-passing metabolism in the rumen. Micellar formation and subsequent metabolic events are identical with those of monogastrics. In the young ruminant the rumen is at first non-functional. Pancreatic lipase activity is high, and triglycerides in the diet entering the small intestine are digested and absorbed as in the non-ruminant. In the pre-ruminant calf, salivary (pre-gastric) lipase released in the saliva acts in the abomasum to cleave butyric acid from milk fat to free fatty acids.

Absorption phase

Fats are absorbed mainly in the proximal small intestine. Their mechanism of absorption has been disputed for a number of years and is thought to occur either by pinocytosis or, more probably, by passive (lipid-soluble) transport.

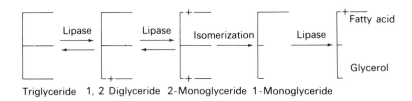

Triglyceride   1, 2 Diglyceride   2-Monoglyceride   1-Monoglyceride

Fig. 14.1   Digestion of fats by pancreatic lipase.

In 1959 S. L. Palay and L. J. Karlin observed that during fat absorption by the small intestine globules of unhydrolyzed triglyceride collected in small vesicles at the bases of the microvilli and were taken into the cell by a process akin to pinocytosis. These workers noted, however, that pinocytotic activity appeared insufficient for the rapid accumulation of fat within the cell. In contrast, J. R. Senior and K. J. Isselbacher in 1960 suggested that the products of triglyceride hydrolysis entered the microvillous membrane by simple (lipid-soluble) diffusion. Moreover, E. W. Strauss and S. Ito (1965) showed that monoglycerides and fatty acids entered the absorptive cells at 0°C, that is, at a temperature when pinocytosis did not occur. R. R. Cardwell, Jr., and co-workers (1967) confirmed that there was no appreciable uptake of fat by pinocytosis, and suggested that the pits in between the microvilli contributed their contents to lysosomes in the apical cytoplasm.

Current evidence thus strongly supports the entry of fats into absorptive cells by passive (lipid-soluble) transport. The hydrolytic products of triglycerides —monoglycerides, fatty acids and glycerol—probably do not enter in free form but link to bile salts as micelles. The micelles first slowly diffuse across the 'unstirred' water layer overlying the glycocalyx, filter through the glycocolyx, and then come into contact with the microvillous membrane where they are disrupted. Bile salts remain in the lumen while monoglycerides, fatty acids and glycerol diffuse through the membrane (Fig. 14.2). Alternatively, the micelle might pass through the membrane intact, but this is unlikely for bile salts are not usually absorbed in the proximal small intestine although limited diffusion is possible. Glycerol is also absorbed by simple (pore) diffusion and carrier-mediated transport.

Once inside the absorptive cell, monoglycerides and fatty acids probably become attached to carrier proteins and are transported by the smooth to the rough endoplasmic reticulum where the enzymatic requirements for the re-synthesis of triglycerides are localized (Fig. 14.3). There, accumulated triglyceride droplets are converted into chylomicrons which are then transported and stored in the Golgi apparatus where possible further synthesis occurs. *Chylomicrons (chylomicra)* are particles, about 1500 Å in diameter, consisting of 88 per cent triglycerides, seven per cent phospholipids, three per cent cholesterol and cholesterol esters, and two per cent protein. The protein is synthesized on ribosomes and combines with free cholesterol and its ester to form a β-lipoprotein coat for the chylomicrons. The phospholipids are derived as intermediates of the glycerophosphate pathway (below) during the cellular metabolic phase. Triglycerides are also incorporated as *very low density lipoprotein*. Saturated, but not unsaturated, fat is transported by such particles, as are some endogenous fatty acids and cholesterol.

Cellular metabolic phase

The synthesis of triglycerides from monoglycerides and fatty acids is complex; hence a simplified outline is given (Fig. 14.4). Synthesis takes place on ribosomes attached to the rough endoplasmic reticulum *(microsomes)* and on the smooth endoplasmic reticulum by two separate pathways.

L-α-GLYCEROPHOSPHATE PATHWAY This pathway is found in most cell systems and involves the esterification of fatty acids to triglycerides. The initial step involves the activation of long-chain fatty acids with coenzyme A, ATP and magnesium ions, to yield fatty acid co-enzyme A, the reaction being catalyzed by the enzyme thiokinase in the microsomes. The fatty acid co-enzyme A then interacts with L-α-glycerophosphate to form phosphatidic acid intermediates. The glycerophosphate is derived mainly from tissue glucose and to a lesser extent from the esterification and phosphorylation of glycerol, catalyzed by glycerokinase in the cell cytoplasm. Diglycerides, formed by dephosphorylation,

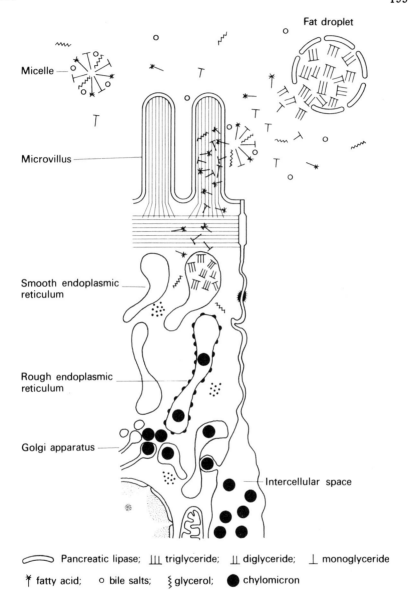

Fig. 14.2   Absorption and transfer of fat in the absorptive cell. Details in text (adapted from Porter, 1969).

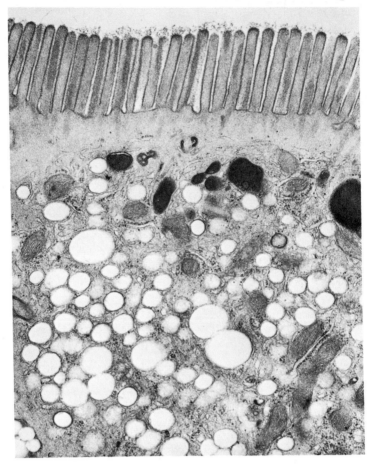

Fig. 14.3   Apical lipid droplets (resynthesized triglyceride) located within the endoplasmic reticulum below the terminal web of the absorptive cell. The enzymes for triglyceride resynthesis are microsomal. The electron-dense bodies are probably lysosome derivatives. Transmission electron microscopy. (Reproduced by permission from Rubin, W. (1971) *Amer. J. clin. Nutr.* **24**, 58.)

are acylated to triglycerides by specific microsomal fatty acetyl co-enzyme A derivatives (monoglyceride and diglyceride acetyl transferase).

MONOGLYCERIDE PATHWAY This pathway is peculiar to the absorptive cell. Monoglycerides are directly acylated by fatty acetyl co-enzyme A derivatives to yield diglycerides and then triglycerides, thus conserving energy. Specific

Fig. 14.4   Digestion, absorption and transport of triglycerides
(adapted from various authors)

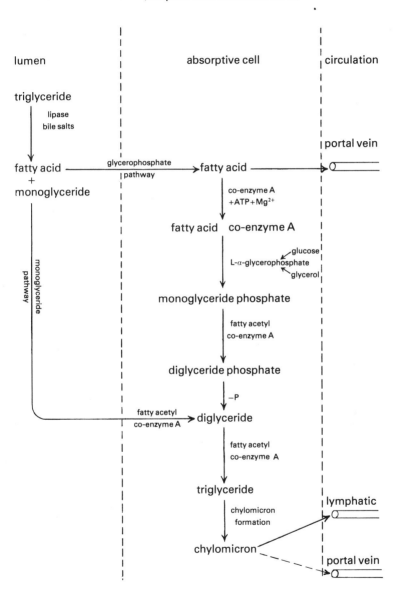

microsomal monoglyceride lipase (or esterase) will hydrolyze any excess monoglycerides when fatty acids are insufficient for the synthesis of triglyceride from monoglyceride.

The relative importance of these metabolic pathways in the absorptive cell varies with different animals. For example, triglycerides are synthesized by both the glycerophosphate and monoglyceride pathways in rats and pigs but the glycerophosphate pathway is more important in sheep. The relative utilization of the two pathways probably depends on the availability of monoglycerides and fatty acids to the absorptive cell. When there is an exces of monoglyceride it is directly acylated by the monoglyceride pathway and hydrolyzed by monoglyceride lipase (esterase); when there is an excess of fatty acids the glycerophosphate pathway is used. When there is an equal amount of monoglyceride and fatty acids the monoglyceride pathway is more important.

Route

Triglycerides, as chylomicrons or very low density lipoprotein, leave the Golgi apparatus and travel by the smooth endoplasmic reticulum to the intercellular spaces of adjacent absorptive cells, probably leaving the cell by 'reverse' pinocytosis (antipinocytosis, excytosis) at the lateral plasma membrane (Fig. 14.2). Since the vesicles bordering the intercellular spaces are often coated while the chylomicrons in the intercellular spaces appear to lack an enveloping membrane, it is possible that the chylomicrons lose their surface coat during the extrusion process from the absorptive cell. Alternatively, the Golgi membrane enveloping the chylomicrons may come into contact and fuse with the lateral plasma membrane, when both membranes become perforated at the point of fusion, releasing the chylomicrons into the intercellular spaces.

Chylomicrons in the extracellular fluid reach the systemic circulation mainly by the lymphatics, although about one-fifth of the total enters the circulation by the portal vein. Medium and short-chain fatty acids do not require chylomicron formation. Medium-chain fatty acids become hydrolyzed after entry by intracellular lipase (esterase) on the microsomes, but are not usually re-esterfied, and leave the cell and enter the bloodstream partly bound to serum albumin and become oxidized in the liver to form complex fats. Short-chain fatty acids pass unesterified into the cell and enter the circulation by the mesenteric veins.

The blood capillary wall, or *endothelium,* consists of a single layer of squamous cells pierced by pores with an average diameter of 100 Å that enable the passage of small molecules by simple (pore) diffusion. Relatively few pinocytotic vesicles, 750–1000 Å in diameter, account for the passage across the capillary of larger molecules (M. W. 10 000–50 000) as, for example, medium and short-chain fatty acids.

Other substances enter the lymphatic circulation by a central, blind-ended lacteal within the lamina propria of the villus. The lacteal endothelium, five to six times thicker than that of the capillaries, shows considerable overlapping and interlocking of epithelial processes which form junctional complexes, leaving small inlet valves 10–40 Å wide, and many prominent intracellular pinocytotic vesicles, with an average diameter of 1600 Å occupying up to 15 per cent of the cytoplasmic volume. Macromolecules (M. W. > 50 000), including chylomicrons, cross the lacteal endothelium mainly in pinocytotic vesicles and are released in the lacteal lumen. Fluid enters at the same time through the vesicles and inlet valves. These valves normally act as safety valves; in diseased tissue they open more frequently. Intestinal peristalsis and contractions of the smooth muscle strands in the lamina propria surrounding the lacteal pulsate the villi, forcing both fluid and chylomicrons down the swollen lacteal lumen and into the lymphatic system.

Cholesterol

Although related to fats cholesterol is very different in chemical structure. Cholesterol in the body is derived both from dietary and non-dietary sources. The normal daily diet in man contains 0.5 g of cholesterol, found only in animal foods, about half of which is absorbed. However, the liver and other regions, including the distal small intestine, synthesize one to three g of both free and esterified cholestrol each day from fats and carbohydrates. Some cholesterol is also derived from desquamated intestinal epithelial cells.

The absorption of cholesterol normally requires the presence of bile salts, long-chain fatty acids and an enzyme, cholesterol ester hydrolase, in the pancreatic juice. Free cholesterol is mainly esterified with fatty acids before being incorporated as micelles, to be slowly absorbed, probably by passive transport, in the proximal small intestine. Intracellular cholesterol also becomes esterified by microsomal cholesterol ester hydrolase. Both esterified fractions then become incorporated in chylomicrons and very low density lipoprotein and are slowly released into the lymphatic circulation. Any free cholesterol escaping into the lymph becomes esterified with fatty acids. Excess cholesterol is removed from the liver as bile acids, mainly cholic, deoxycholic and taurocholic acids. The acids are neutralized by cations, sodium and potassium and constitute the bile salts.

Bile salts

Bile salts are synthesized in the liver by cholesterol. They are excreted from the liver by the bilary tract into the duodenal lumen and eventually rapidly absorbed by the ileum into the portal vein and then re-excreted by the liver (enterohepatic circulation; Chapter 5). Bile salts are re-circulated about eight times each day, an insignificant amount being lost in the faeces and urine.

In man, about 25 g of bile salts are reabsorbed daily. Bile salts are conjugated before being secreted into bile. Intestinal bacterial enzymes first hydrolyze con-

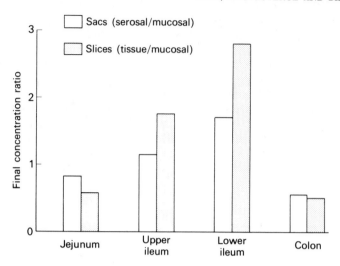

Fig. 14.5   Absorption of bile salt (radioactive cholic acid) by segments of rat small and large intestine. A concentration ratio greater than 1.0 indicates absorption (adapted from Holt, 1964).

jugated bile acids to unconjugated (free) bile acids. For example, cholic acid is converted to deoxycholic acid, which is absorbed and returned to the liver in the portal vein partly bound to serum albumin, to be conjugated with glycine or taurine and re-cycled.

Absorption of bile salts has been studied in rats by P. R. Holt (1964) using various *in vitro* preparations. The absorption of one bile salt competitively inhibits the absorption of another bile salt, suggesting that there is a common receptor site on the microvillous membrane. Transport is abolished by different metabolic inhibitors, oxygen deprivation or low temperature and exhibits saturation kinetics, showing that the process is an active one. Absorption is reduced either at low concentrations of sodium or when sodium is replaced by potassium, lithium, choline or Tris. Although the active transfer process requires the presence of sodium ions, it is not associated with the generation of an electrochemical gradient. Thus, the active transport of bile salts is similar in most respects to the energy-dependent system for active transfer of sugars and amino acids.

Bile salts are actively absorbed in the ileum, particularly distally (Fig. 14.5). This site of absorption in different mammals appears to be unique in studies on active absorption of different solutes. However, limited passive transport in some species is also likely throughout the small and large intestines. Although vitamin B_{12} is also absorbed in the ileum, the absorptive process is possibly by pinocytosis.

15   Water

The epithelia in certain regions of the alimentary system secrete large amounts of fluid. In man, about eight litres of secretions from the saliva, gastric juice, bile, pancreatic juice and small and large intestinal juice are all poured each day into the alimentary canal (Fig. 15.1). Together with a normal daily intake of about one and a half litres of fluid, well over nine litres of isotonic fluid are reabsorbed, mainly in the small intestine and to a lesser extent in the large intestine, thereby completing an efficient enterosystemic cycle.

Water filters passively in and out through the microvillous membrane pores of the absorptive cells and possibly through the tight junctions in between adjacent cells, in response to osmotic pressure gradients created by the transmucosal movements of solutes, until the osmolarity in the lumen and plasma are equal. Thus, the process of water absorption and secretion is closely linked, absorption being the net increase over inward and outward movements. There is no net absorption of water in the duodenum but about half the total volume of fluid entering the small intestine is absorbed in the jejunum and a further third in the ileum. The jejunum is therefore more readily permeable to osmotically induced water transfer than in the ileum, probably because the membrane pores are larger, thus allowing rapid adjustments of osmotic pressure arising secondary to the more effective transport of solutes in this region (p. 83).

Progress in understanding the mechanism of water absorption has fallen well behind the number of investigations made. Moreover, only few experiments have been attempted using the small intestine, and hence the process of water absorption in the small intestine is inferred mainly from work done either on models or on other epithelia. Several hypotheses have been suggested to explain the process of water absorption, some plausible, others not. What follows summarizes the main mechanisms that have been proposed.

1 Filtration

Filtration plays a minor role in the absorption of water since the hydrostatic pressure in the lumen of the intestine is too low to account for the filtration of water molecules under pressure into the absorptive cell. Experiments have shown that a filtration force or hydrostatic pressure far above that normally found in the lumen of the small intestine had little effect in water absorption.

2 Classical or 'exogenous' osmosis

Passive inward water movement may cross a biological membrane from a region

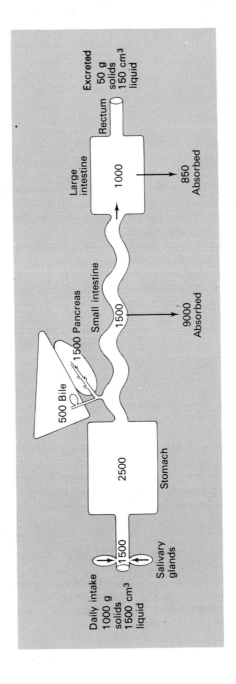

Fig. 15.1   Secretions (cm³) of the alimentary canal in man. Over eight litres of fluid are secreted each 24 h which, together with the daily fluid intake, are mainly absorbed in the small intestine.

of low concentration (hypotonic) into a region of higher concentration (hypertonic) by classical osmosis. The sodium pump actively transports solutes and in doing so may create an osmotic pressure gradient down which water will passively flow and accumulate in the cell. However, absorptive cells are isosmotic with the blood (300 mmol 1^{-1}), and so water movement from the intestinal lumen into the cells by osmosis is unlikely. Moreover, the gall bladder epithelium can transport water against an osmotic gradient, and hence water absorption by osmosis seems improbable.

3 Electro-osmosis

As the plasma membrane of the absorptive cell has a small electrical potential difference across its surface it is possible that water transfer occurs by electro-osmosis. But experiments have shown that potentials ten times or more than those actually measured are required for water absorption to occur by this process.

4 Pinocytosis

Calculations have shown that, in order for the small intestine to absorb its daily water load by pinocytosis, the absorptive cell would require to make pinocytotic vesicles ten times the rate normally seen. Hence, water absorption by pinocytosis cannot occur to any significant extent.

5 Co-diffusion

When solute molecules passively diffuse through aquatic pores in the apical cell membrane water molecules might be swept along with the solute; this is co-diffusion, or frictional drag. This process can cause water to move 'uphill' against its concentration gradient. However, water entry by this route is not likely to be significant since the transported fluid in the absorptive cells is usually isotonic with the fluid in the lumen.

6 Local or 'endogenous' osmosis

Inward water movement by one or more of the mechanisms outlined above is therefore unlikely or unimportant. Several attractive working models have been proposed to show that water absorption by osmosis *within* the cell is possible. The local or 'endogenous' osmosis hypothesis states that active transport of solutes creates a local osmotic gradient either inside or in close proximity to the cell which causes an inward flow of water by osmosis.

Durbin (1960) suggested that the plasma membrane is a compound one, being pierced by small pores in series with larger pores. The small pores each have a radius of at least 4 Å which allows the entry of sodium, chloride, and water molecules, which have crystal radii of 0.97 Å, 1.81 Å and 1.40 Å respectively. Sodium ions pass through the outer small pores and become concentrated within the larger pore compartments, causing a transient increase in hydrostatic pressure. As a result of this local, high hydrostatic force water molecules passively flow through the small pores into the compartment, to be forced through the larger pores into the cell. The water molecules then flow through the cell and diffuse out through the basal or lateral membrane by a 'downhill' process while sodium ions are pumped out. The compound membrane thus allows apparent 'uphill' movement of water molecules against a concentration gradient, although water transport is a passive process.

P. F. Curran and J. R. MacIntosh (1962) explained inward water movement by using a working model consisting of a Perspex chamber divided into three compartments (A, B, C) by a cellophane dialysis membrane and a sintered glass disc partition (Fig. 15.2). Different concentrations of sucrose solutions were put in A, the sucrose concentration in B was kept constant, and water put into C. The fluids in compartments A and C were gently stirred. Any change in volume in C, was measured by a graduated pipette fitted to the side of the Perspex chamber at that end.

As long as the sucrose concentration in B was greater than in A, water moved into C (Fig. 15.2). Thus, water apparently moved 'uphill' against a concentration gradient from A to C in the absence of an 'active transport system' in the partition between A and B. This unexpected phenomenon was explained by referring to the osmotic and hydrostatic properties of the partitions. The cellophane membrane has a low permeability for sucrose while the sintered glass disc partition is freely permeable to the sugar. The entry of water from A into B is governed by the osmotic pressure gradient across the cellophane partition; this generates a hydrostatic pressure in B which drives water across the sintered glass disc partition into C. The net result is that water flows from A through B into C, even though the osmolarity in A is greater than in C.

In 1966 J. M. Diamond and J. McD. Tormey adapted this model for fluid movement through the gall bladder epithelium. Ultrastructure studies showed that when the cells are absorbing fluid from the bile in the liver the intercellular spaces are 1 μm thick and only 0.1 μm thick when not, indicating that the main route of fluid movement through the cells is through the lateral plasma membrane. Swollen intercellular spaces have also been observed in the transporting epithelia of both the small and large intestines. These workers suggested that sodium pumps are situated along the upper margins of the plasma membrane adjacent to the intercellular spaces, while water diffuses through the lower margins of the lateral membrane (Fig. 15.3a). The entry of sodium and water into the intercellular channels creates a hypertonic solution which becomes diluted and isotonic with the blood as it diffuses through the basement membrane and underlying sub-epithelium. There is thus a standing osmotic gradient in the intercellular spaces. The resulting increased hydrostatic pressure forces the fluid into the lacteals or blood capillaries. P. F. Curran (1968) proposed that the

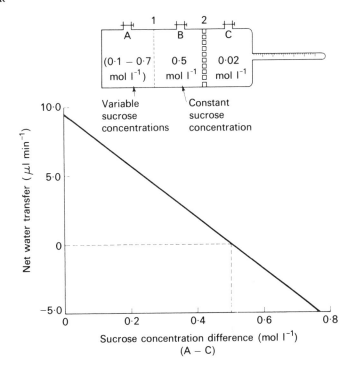

Fig. 15.2   Water flow in an artificial model. A three-compartment Perspex box was sub-divided by a cellophane dialysis membrane (1) and a sintered glass disc partition (2). Various concentrations of sucrose solution were put into compartment A in different experiments while the concentration of sucrose in compartment B was kept constant. Changes in the volume of water in compartment C were measured with a graduated pipette. As long as the sucrose concentration in compartment B was higher than in compartment A, there was a net volume flow from compartment A through B and into C (modified from Curran, 1965, after Curran and MacIntosh, 1962).

sodium pumps were equally distributed along the length of the lateral plasma membrane adjacent to the intercellular spaces and that the entry of sodium and water molecules created an unstirred, hypertonic, fluid-filled region (Fig. 15.3b). By analogy, the interior of the absorptive cell represents the first compartment (A) proposed earlier by Curran and MacIntosh (Fig. 15.2), the lateral plasma membrane is the cellophane membrane (1) and the intercellular space the second compartment (B). The basement membrane and lacteal or capillary wall represent the freely permeable sintered glass disc partition (2), and the lumen of the lacteals or blood capillaries represents the third compartment (C).

WATER

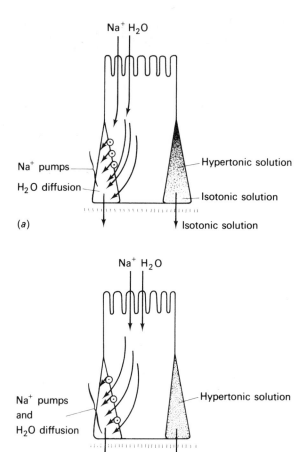

Fig. 15.3   Proposed models for 'local' osmosis of water entry into the absorptive cell. Details in text. ((a) after Diamond and Tormey, 1966 and (b) Curran, 1968).

A somewhat similar model for the transport of sodium chloride and water in the rumen epithelium has also been proposed (Fig. 15.4). This epithelium is multi-layered, with dead cells being extruded into the lumen by the development of cells beneath. Sodium chloride and water enter the epithelium by diffusing into keratinized (dead) cells. Deeper in the epithelium sodium chloride molecules are pumped into intercellular channels, creating a concentrated standing osmotic gradient which becomes diluted as water molecules diffuse downwards. Back-

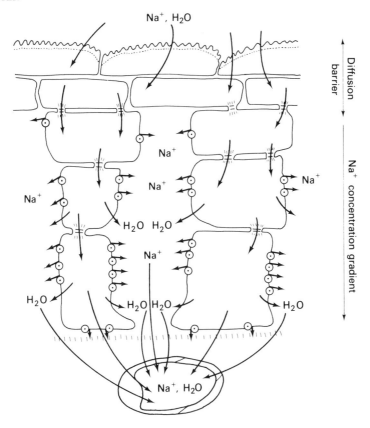

Fig. 15.4   Sodium and water entry into local intercellular compartments of the rumen epithelium. Details in text (after Steven and Marshall in Phillipson, 1970).

diffusion is prevented by the keratinized layers. This mechanism has received support from histochemical studies in which membrane ATPase has been located along the lateral plasma membrane of the non-keratinized cells.

7 *Villous counter-current exchange hypothesis*

Finally, H. Haljamäe and colleagues (1973) have recently provided an attractive alternative to account for fluid transfer in the villi, based partly on endogenous osmosis as described above and partly on the well-known counter-current exchange and multiplication system operating in the mammalian kidney. Using perfused *in vivo* cat jejunal segments, autoradiographic and tissue sectioning

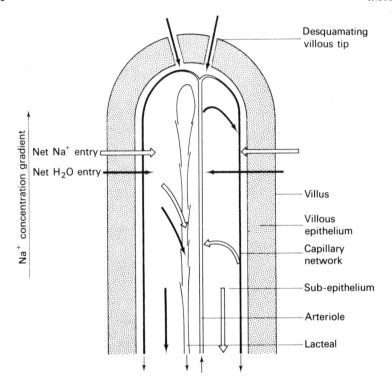

Fig. 15.5   Extravascular shunting of sodium and water in the villous core, accor-
ding to the counter-current exchange hypothesis. Details in text (adapted from
Haljamäe *et al.*, 1973).

techniques the sodium concentration in the lamina propria was found to be three
to four times higher at the villous tip than at its base. The steep sodium gradient
observed along the villous length is believed to be maintained by a counter-
current multiplication of sodium taking place at the hairpin vascular loop formed
by the unbranched central arteriole and the dense capillary network (Fig. 15.5).
Therefore, the creation of a sodium concentration gradient in the villous
sub-epithelium depends on blood flow. The actively absorbed sodium is short-
circuited from the capillary network to the central arteriole, resulting in re-
circulation and concentration towards the villous tip. The increased sodium concentration and osmolarity in the capillaries causes transfer of water from the
central arteriole to the capillary bed, and sodium becomes further concentrated.
The hypertonicity in the villous core creates an increased hydrostatic pressure in
this region, forcing fluid through inlet valves and vesicles into and down the central lacteal (over). The villous core represents the hypertonic compartment (B)

outlined above, and the vascular counter-current exchange system and/or the lacteal represents the sintered glass disc partition (2). The addition of glucose to the luminal fluid significantly increased the sodium concentration along the villous length, while intra-arterial infusion of ouabain greatly reduced the sodium concentration.

In conclusion, of the various mechanisms which have been proposed to explain water transfer into the villus, either local ('endogenous') osmosis or the villous sub-epithelial counter-current exchange hypothesis, or a combination of both, appears to be acceptable on the basis of information currently available.

Route

Under *in vitro* conditions J. E. Lee from 1961 showed that fluid transferred across the rat jejunal wall flowed mainly into the lymphatic ducts and not into the mesenteric veins. The lymphatic flow accounted for 85 per cent of the total volume of fluid transported, the venous flow 11 per cent and the serosal flow four per cent. Initially, water and electrolytes appear to be mainly absorbed into the lymphatic system to be secondarily transferred to the venous blood capillaries by an unknown pathway. Recent work has confirmed that during absorption net gain of water in the circulation is exclusively by the lymphatic system. Only when there is an osmotic gradient between blood and intestinal lumen is water initially transferred by the intestinal capillaries. Hence, the formation of lymph in the intestine is closely associated with water and solute absorption.

Energy supply for fluid transfer

Fluid absorption in the small intestine is greatly increased by the presence of glucose in the lumen and virtually stops when phloridzin or metabolic poisons are added. Fluid transport in the presence of galactose, fructose or different amino acids is much lower. There appears to be two separate mechanisms for fluid transport by the small intestine. In 1961 B. A. Barry and co-workers first showed that fluid transfer in the rat jejunum depends on the presence of luminal glucose, but in the ileum it does not. Isolated segments from everted small intestine (excluding the duodenum) were incubated either with or without glucose and fluid transfer in each segment determined. Glucose-dependent fluid transfer was more important proximally and glucose-independent transfer was more important distally. The total fluid transferred was greatest in the mid-intestine. Both mechanisms required the presence and active transport of sodium. It is not known whether the mechanisms function on the basis of local osmosis.

By incubating intestinal segments with specific metabolic inhibitors the source of energy for fluid transfer can be determined. Fluoride inhibits the glycolytic cycle and fluoroacetate inhibits the tricarboxylic acid cycle (p. 114). Using rat

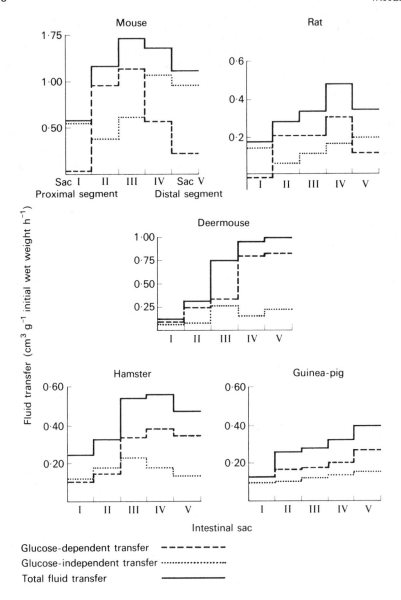

Glucose-dependent transfer   – – – – – – –

Glucose-independent transfer ⋯⋯⋯⋯⋯⋯⋯

Total fluid transfer   ——————

Fig. 15.6 Glucose-dependent, glucose-independent and total fluid transfer from different *in vitro* segments of equal length of the entire small intestine in laboratory animals. Sac I: proximal segment; sac V: distal segment. Details in text (after Madge, 1972).

everted intestinal segments J. K. Detheridge and colleagues (1966) found that the presence of fluoride in the mucosal fluid inhibited the glucose-dependent system while the presence of fluoroacetate inhibited the glucose-independent system. Thus, the energy for glucose-dependent fluid transfer in the proximal rat intestine depends mainly on the glycolytic pathway and is sustained by an exogenous energy supply, while the glucose-independent fluid transfer distally depends mainly on the tricarboxylic acid cycle and is sustained by an endogenous energy supply.

Figure 15.6 compares the energy supply for fluid transfer in the small intestine of several laboratory animals. In all animals studied, the stimulation of fluid in the presence of D-glucose was approximately twice that in its absence. In the mouse and rat, glucose-dependent fluid transfer occurred mainly in the mid-intestine, glucose-independent fluid transfer occurred mainly in the distal intestine, while total fluid transfer was greatest in the mid-intestine. The energy required for fluid transfer in the proximal small intestine was presumably provided by the glycolytic cycle while the energy required for fluid transfer distally was presumably provided by the tricarboxylic acid cycle. By contrast, in the deer-mouse, hamster and guinea-pig, glucose-dependent transfer was usually higher than glucose-independent transfer throughout the small intestinal length, especially distally, while total fluid transfer generally increased from the proximal to the distal intestine. In these animals the energy required for fluid transfer by the intestine was presumably mainly provided by the glycolytic cycle. Thus, the energy required for fluid transfer is species specific and generalizations are misleading. The capacity for total fluid transfer also varied with the different species. For example, in the mouse and deermouse small intestine about three times more fluid was transferred compared with the rat, hamster and guinea-pig small intestine. Hence, fluid utilization from both water-intake and endogenous secretions also appears to depend on the animal studied.

Fluid transport along the small intestine is also associated with an electro-chemical gradient. Thus, in the rat jejunum fluid transfer is energized by anaerobic metabolism and stimulated by metabolizable sugars, but is not asociated with an increase in the electrical potential difference, that is, a non-electrogenic sodium pump exists for the fluid transfer mechanism. By contrast, in the ileum, fluid transport is energized by aerobic metabolism but not stimulated by metabolizable sugars, and is accompanied by a rise in the electrical potential difference, that is, an electrogenic sodium pump is required for the fluid transfer mechanism.

16 Electrolytes and minerals

Electrolytes

The body water contains electrolytes dissolved in different proportions. Electrolytes are substances composed of elements which can be electrolytically dissociated into anions and cations. In the extracellular fluid (in man, tissue fluid, eleven litres and blood plasma, three litres) sodium is the main cation and chloride and bicarbonate are the main anions. The intracellular (cell) fluid (35 litres) contains potassium and magnesium as the principal cations and only a low concentration of sodium ions, while the chief anions inside the cell are phosphate and protein. In the secretions of the alimentary tract the main cation is sodium except in the gastric juice where it is hydrogen. The main anion in the secretions is chloride except in the pancreatic and large intestinal juices where it is bicarbonate (Fig. 16.1).

Electrolytes in the alimentary canal originate from both the food intake and secretions of the alimentary tract. The composition of electrolytes and water in the lumen represents the simultaneous fluxes from lumen to blood and from blood to lumen. Therefore, absorption is the net increase and secretion the net decrease in electrolytes and water resulting from bidirectional fluxes in and out of the lumen. Electrolytes and water are absorbed mainly by the small intestine into the circulation, reach the kidneys where they are largely re-absorbed and returned to the circulation, to be distributed in the body water.

Hormones can act in regulating electrolyte and water movements across the small intestinal mucosal surface. For example, aldosterone, a hormone of the adrenal cortex which promotes the re-absorption of sodium from the kidney tubules, has a slow but stimulating effect on *in vivo* but not *in vitro* absorption of sodium, chloride and water in the ileum and large intestine but not in the jejunum. The effect of this hormone on potassium is uncertain, although possibly associated with secretion. Aldosterone probably promotes the synthesis of both ATP and membrane ATPase and hence increases the efficiency of the sodium pump. Many gastrointestinal hormones can also affect absorptive or secretory activities in the small intestine. Thus, secretin and CCK–PZ decrease sodium, potassium, chloride and water absorption; secretin does not affect bicarbonate exchange but CCK–PZ causes bicarbonate secretion. Gastrin inhibits electrolyte and fluid transfer in the ileum but not in the duodenum or jejunum. Although 'gastric inhibitory polypeptide' and 'vasoactive intestinal polypeptide' stimulate water secretion they do not affect electrolyte composition.

Electrolyte and water exchanges between the human intestine and the in-

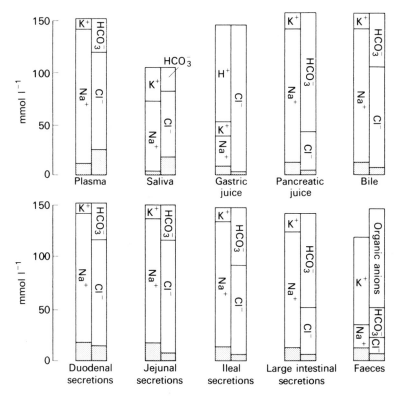

Fig. 16.1 Electrolyte composition of secretions of the mammalian alimentary tract compared with that of plasma. Shaded areas include cations (Ca^{2+}, Mg^{2+}) and anions (HPO_4^{2-}, SO_4^{2-}) (after various authors).

testinal lumen are complex and not fully understood. In the *duodenum* there is little or no net absorption of water, electrolytes or solutes from isotonic solutions. By contrast, the *jejunum* avidly absorbs water and electrolytes against a small electrochemical gradient. The presence of solutes (sugars or amino acids) promotes marked increased absorption of water and sodium. The active absorption of sodium is coupled with that of both chloride and bicarbonate; other sodium ions are absorbed in exchange for hydrogen ions. Potassium absorption is passive. Water absorption is also passive and consequent upon net movement of solutes from the lumen. Unlike the jejunum, absorption of water and sodium in the *ileum* and *large intestine* is slower and independent to that of solutes. These movements take place against a steep electrochemical gradient. Sodium is actively absorbed in the ileum in exchange for hydrogen and accompanied by some chloride, while potassium is passively absorbed in the ileum and secreted in the large intestine. The rapid absorption of chloride is balanced by the secretion of bicarbonate. In the *faecal fluid* the main cation is potassium and the sodium concentration is low, while organic anions derived from bacterial activity replace chloride as the main anion.

The rumen in digastric animals performs several functions also carried out by the ileum and large intestine. For example, the rumen absorbs sodium and chloride secreted in the saliva and releases potassium and bicarbonate, the latter serving to buffer VFA in the lumen.

Sodium

The adult human body contains about 100 g of sodium. The main source of sodium is the sodium chloride in the diet. The normal daily intake of sodium chloride is about ten g, most of which is readily absorbed and distributed in the extracellular fluid and deposited in bone. Between 50 to 200 mg are lost daily by sweating. Sodium chloride is also lost in the urine and faeces.

The unique physiological activities of sodium during absorption have already been given. In the presence of isotonic saline in the intestinal lumen not only sodium, but also chloride and water, are all absorbed. In the absence of isotonic saline, sodium, chloride, and water are secreted. Sodium enters the absorptive cell down an electrochemical gradient and therefore entry probably does not require expenditure of energy. Entry is by a carrier-mediated, facilitated diffusion process. Sodium is extruded from the cell against an electrochemical gradient by a sodium pump located in the lateral or basal plasma membrane. This mechanism uses metabolic energy, catalized by membrane ATPase.

The sodium pump is abolished by ouabain, when intracellular sodium increases as sodium extrusion stops. The pumping mechanism virtually ceases in the absence of sugars or amino acids, or in the presence of metabolic poisons; lack of oxygen or low temperature also stops the pump. Sodium, particularly anteriorly, acts as a co-substrate for the transport of many solutes which is associated, particularly distally, with the prompt increase in a positive electrical potential difference of 10 to 15 mV across the absorptive cell. Some diffusion of sodium ions is likely from the intercellular spaces, in between apposed lateral plasma membranes of adjacent cells, across the tight junctions and into the intestinal lumen. Secretion is also possible through the microvillous membrane.

The net absorption of sodium is equal to the difference between the influx of sodium from lumen to blood and its efflux from blood to lumen. This fact was shown by using radioactive sodium, initially present in either the intestinal lumen or in the blood. When *in vivo* intestinal loops in dogs were filled with isotonic sodium chloride solution containing sodium tracer, the original sodium concentration remained unchanged but the labelled sodium in the lumen rapidly disappeared. Since the total sodium concentration remained constant, a quantity of sodium equal to that leaving the lumen must have entered from the blood. Alternatively, when labelled sodium was injected intravenously and the intestinal loop filled with isotonic sodium sulphate solution, the total luminal sodium concentration fell slightly and the sodium in the lumen was rapidly replaced by radioactive sodium from the blood. The fall in sodium concentration in the lumen indicates net sodium absorption which represents the difference between the sodium influx and efflux.

There is little absorption or secretion of sodium in the duodenum. Net absorption occurs in the jejunum in close association with the transport mechanisms for sugars, amino acids and bicarbonate. In the ileum sodium is absorbed in the absence of solutes. Most of the remaining unabsorbed sodium and water in the small intestine are absorbed by the large intestine and rectum. Although the large intestine has a marked capacity for sodium absorption the sodium load there is low and hence little absorption normally takes place. The large intestine has a sodium pump mechanism which generates a large positive electrochemical gradient of 30 to 40 mV. However, the addition of solutes in the lumen has no effect on sodium or water transfer and so, unlike the small intestine, the sodium-solute coupling mechanism is missing.

Chloride

Chloride is the principal anion of the extracellular fluid. Chloride is actively secreted by the gastric mucosa, actively absorbed by the rumen epithelium and passively absorbed by the large intestine. Although chloride absorption in the small intestine is more rapid in the ileum than in the jejunum, its permeability is half that of sodium. Excess chloride absorption in the ileum and large intestine is balanced by bicarbonate secretion. Chloride is absorbed in the presence of low bicarbonate concentration while at high bicarbonate concentration chloride is secreted.

The mechanism of chloride absorption in the small intestine remains enigmatic. Some investigators have claimed that chloride ions enter the absorptive cells by simple (pore) diffusion as a passive consequence of active sodium absorption. More recently, others have indicated that chloride absorption moves 'uphill' against an electrochemical gradient, obeys the saturation kinetics law, is inhibited by metabolic inhibitors (cyanide, dinitrophenol) and by other anions (bromide, iodide, nitrate). Thus, chloride absorption appears to involve an energy-dependent, carrier-mediated, active transport mechanism. Chloride exit from the cell through the basal or lateral membrane is by simple or carrier-mediated diffusion down an electrochemical gradient. Secretion of chloride ions from the intercellular spaces, in between adjacent lateral plasma membranes and across tight junctions into the intestinal lumen is also possible.

Chloride transport, therefore, seems to be the mirror image of that proposed for sodium transport. Sodium absorption through the apical membrane is by carrier-mediated facilitated diffusion down an electrochemical potential gradient, and its removal through the basal or lateral membrane is an active transport process against an electrochemical gradient. Chloride entry is by an active process through the apical border against an electrochemical gradient, and its removal at the basal or lateral membrane is by simple or facilitated diffusion down an electrochemical gradient. Although both sodium and chloride are actively absorbed by the intestine, diametrically different mechanisms are thus involved.

Potassium

The adult human body contains about 150 g of potassium. The normal intake in the food is two to four g per day. Potassium is the main cation of absorptive and other cells and plays an important role in the functioning of nerve and muscle cells within the intestinal wall. Potassium ions move in either direction across the entire length of the small intestinal epithelium, although at a much slower rate than that of sodium ions. When the potassium concentration in the lumen is lower than that of the blood, potassium is secreted, and when it is higher it is absorbed. Normally, the potassium concentration in the lumen is such that there is a small net absorption.

Each day an adult absorbs only two to three per cent of the total exchangeable potassium. Recent work indicates that the epithelial surface is relatively impermeable to potassium, bidirectional fluxes taking place through tight junctions in between adjacent epithelial cells; there is little mixing with intracellular potassium.

Water absorption is associated with potassium absorption and water secretion with potassium secretion, and so potassium, like water, is probably absorbed by simple (pore) diffusion. Furthermore, there is a linear relationship between net transport of potassium and its luminal concentration.

The transport of glucose requires the presence of potassium ions inside the absorptive cells because potassium is required during intracellular metabolism of glucose. Moreover, some intracellular enzymes (fructokinase, hexokinase) are activated by potassium and inhibited by sodium. Also, potassium ions are required for normal membrane ATPase and membrane disaccharidase activities. Finally, the presence of potassium in the intestinal lumen is necessary for maintaining sodium absorption to its fullest extent.

Although there is evidence to show that the limited potassium absorption in the large intestine is an active process, potassium is nevertheless normally secreted into the lumen. A coupling mechanism between sodium and potassium in the large intestine has been suggested, for a decrease in sodium absorption is associated with an increase in potassium absorption, and an increase in sodium is associated with a decrease in potassium.

Bicarbonate

Bicarbonate and chloride are the main anions of the extracellular fluid. The human body contains about 45 g of bicarbonate, most of which is formed endogenously by the hydration of carbon dioxide. Pancreatic juice and bile produce high concentrations of bicarbonate which are poured into the lumen of the duodenum. Some sodium ions are actively absorbed in the jejunum in exchange for hydrogen ions. In the lumen hydrogen ions convert some bicarbonate ions to carbon dioxide, tending to make the jejunal contents acid (pH 5–7). Residual bicarbonate ions in the jejunum are rapidly absorbed by an active process. In the ileum bicarbonate ions are actively secreted into the lumen in exchange for some chloride ions, tending to make the ileal contents more alkaline (pH 6.5–8.0). Lack of bicarbonate in the intestinal lumen decreases the absorption of non-electrolytes, electrolytes and water.

Calcium

The human body contains about 1300 g of calcium, most of which is stored in bone and teeth and the remainder found in the plasma. Calcium plays an essential role in diverse physiological processes. The normal intake in the adult human diet is about 800 mg each day, of which half or less is absorbed and the rest excreted, mainly in the faeces. Calcium is also secreted by the small intestinal epithelium so that faecal calcium often exceeds ingested calcium. Calcium transport involves a rapid entry mechanism and a slow exit mechanism. At low calcium concentrations entry into the cell is by facilitated, carrier-mediated transport; at high concentrations absorption is also by simple (pore) diffusion. Calcium-binding proteins may act as carriers within the microvillous membrane, or serve as calcium absorptive sites, or concentrate calcium at the apical surface. The role of sodium in calcium transfer is inconclusive for calcium uptake does not entirely depend on luminal sodium. The exit mechanism at either high or low calcium concentrations is stopped by lack of sodium, oxygen deprivation or low temperature, and moves against both chemical and electrochemical gradients, indicating that it is an active transport process, possibly mediated through basal membrane calcium ATPase. Calcium absorption is also linked to cell metabolism. This is because actively transported, non-metabolizable sugars in the lumen inhibit calcium transport. Metabolizable sugars that are not actively transported, and also those that are both transported and metabolized, increase calcium absorption. Cellular calcium uptake involves active transport of calcium into mitochondria.

Vitamin D controls the activity of calcium-sensitive ATPase in the microvillous membrane and hence calcium uptake. Vitamin D also induces the formation of calcium-binding proteins located on the surface of the microvilli; their role in calcium translocation is unknown. Lack of vitamin D in the diet depresses the rate of calcium absorption; feeding vitamin D to deficient animals increases calcium absorption. The mechanism controlling this action is unknown.

The physical state of calcium will also influence the absorption rate. Thus, soluble calcium is rapidly absorbed whereas insoluble calcium is poorly absorbed. Calcium absorption may be related to that of magnesium in some animals since magnesium inhibits calcium absorption although calcium does not stop magnesium uptake.

Most adult mammals (e.g., rat, mouse, guinea-pig, rabbit) on a normal diet absorb calcium mainly in the duodenum and secrete it from the jejunum and ileum; a few (e.g., dog, hamster) absorb it mainly in the ileum. Calcium secretion in the ileum of adult rats is about twice that in the duodenum and apparently involves an active transport process. New-born rodents absorb calcium mainly in the ileum. However, when they are two to three weeks old they absorb it mainly in the duodenum, like the adults. The efficiency of calcium absorption decreases with increasing age. Pregnant or lactating animals absorb calcium much faster than non-pregnant ones. Animals deprived of calcium in their diet can absorb extra calcium from the entire length of the small intestine, possibly by passive (pore) diffusion. The mechanism mediating this response is unknown.

Phosphorus

The human body contains about 800 g of phosphorus, four-fifths of which is combined with calcium in the bones and teeth as phosphate and the rest widely distributed in the body, mainly as a constituent of ATP. The dietary intake in man is about 800 mg daily, two-thirds of which is absorbed in the form of inorganic phosphate in the proximal small intestine and the rest excreted in the urine. In ruminants, phosphate is mainly absorbed in the rumen and the remainder excreted in the faeces.

Phosphorus enters the absorptive cell against a concentration gradient and leaves it down a concentration gradient, and experiments have suggested that active transport and facilitated diffusion respectively are involved. Although an increase in phosphorus absorption is accompanied by a decrease in calcium absorption, and a decrease in phosphorus by an increase in calcium, phosphous absorption is independent of calcium transport. Vitamin D is normally required for the absorption of phosphorus.

Magnesium

Over half of the total body magnesium of 25 g in man is located in bone and most of the remainder found in the intracellular fluid. In humans less than half the daily intake of 300–350 mg is absorbed. Both the site of magnesium absorption and the amount absorbed depend on the species studied. In man, rats and rabbits magnesium is slowly absorbed by both the small and large intestines. In young rats, although more is absorbed in the ileum than in the jejunum, about half of the total dietary magnesium is absorbed in the caecum and large intestine. In calves, magnesium is absorbed both in the small and large intestines but as the animal matures the large intestine apparantly loses its capacity for magnesium absorption.

The slow transfer of magnesium in adults suggests a passive transport mechanism, although an active process has also been implicated. The main regulator of magnesium absorption is the amount of magnesium in the intestinal lumen rather than the nutritional requirements of the animal: magnesium absorption is not increased in animals with magnesium deficiency.

Sulphur

Sulphur is an important constituent of cell proteins and therefore widely distributed in the body. It forms part of the molecule of three sulphur-containing amino acids, cysteine, cystine and methionine, and is also a component of other organic compounds, for example, coenzyme A, heparin, insulin and taurocholic acid.

Other Minerals

Minerals which are known to be essential and which are present in the body in relatively large amounts *(bulk elements)* include sodium, potassium, calcium, phosphorus, magnesium and sulphur, and constitute up to three-quarters of all the organic materials in the body. Other essential minerals are required in only very small amounts *(trace elements)* and include iron, cobalt, copper, iodine, manganese, molybdenum and zinc. The distribution, transport and functions of some of these minerals have already been given; those of the remainder are summarized below.

Iron

The human body contains four to five g of iron, most of which is stored in the small intestinal mucosa, bound in plasma, and forms part of the red blood corpuscle (haemoglobin) and striated muscle (myoglobin). The average dietary intake in man is about 15 mg each day, but only about one-tenth of this amount is absorbed, the rest being excreted in the faeces.

Iron is absorbed in two phases: rapidly (and mainly) in the proximal small intestine and slowly in the distal intestine. At low concentrations iron is rapidly absorbed and transferred through the absorptive cell, probably by intracellular carriers; at higher concentrations it is temporarily stored and removed more slowly. Animals which are deficient in iron are able to absorb it more efficiently than normal ones. Animals with an excess of dietary iron absorb it less efficiently.

Inorganic iron uptake and transfer by the small intestine is complex and remains unsolved (Fig. 16.2). Iron is absorbed equally well in the ferrous and ferric states provided that it remains in the ionized form. Possibly iron binds to *gastroferrin*, a mucopolysaccharide iron-binder in the gastric juice, to form an in-

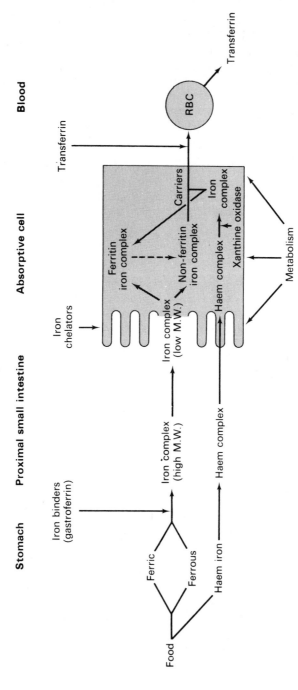

Fig. 16.2 Reactions of dietary iron in the lumen of the stomach and small intestine, and iron transfer across the absorptive cell (tentative).

traluminal transfer system of high molecular weight. Iron chelators (ascorbic acid, some simple sugars, and amino and organic acids), adsorbed onto binding sites on the membrane surface or located within the glycocalyx of the absorptive cell, detach the gastroferrin and form chelated iron complexes of low molecular weight which then enter the cell, possibly by an active transport process.

Following absorption, some of the iron complex appears in the apical endoplasmic reticulum as membrane-bound, electron-dense particles, 50 Å in diameter (F-bodies), containing *ferritin,* forming an intracellular iron pool. Only a small amount of iron in this stored form is transferred to the plasma; the bulk of it (sequestered iron) is released by shedding of the villous cells at the end of their life-span. The amount of intracellular ferritin probably controls the amount of iron uptake and release. However, most of the absorbed iron does not follow the ferritin pathway. *Non-ferritin* iron complex is transferred across the cell, probably attached to carrier proteins, and then slowly but constantly removed from the basal membrane by an unknown but energy-requiring process. Free iron in either the ferric or ferrous state does not exist in the cell. The various stages of iron translocation probably rely on cell metabolism.

As iron molecules enter the circulation they become attached to specific iron-binding proteins, *transferrin* (β-globulin), to be translocated to red blood corpuscles. Once attached to the corpuscle surface, the iron is released into the cell and transferrin becomes detached. A smaller fraction becomes associated with striated muscle.

Iron in meat *(haem iron)* is released as a haem complex in the small intestine by duodenal secretions and absorbed intact by an unknown pathway. It is then split by an intracellular enzyme, xanthine oxidase. Thereafter, the released iron complex follows the same pathways as those which entered the absorptive cell in an ionizable form. Although alternative hypotheses for iron transfer across absorptive cells have been suggested, the one outlined above follows current opinion.

Cobalt

The adult human body contains 15 mg of cobalt and the daily diet contains about two mg. Cobalt is a constituent of vitamin B_{12}. Only about one-fifth of the dietary cobalt is absorbed by passive transport and the remainder excreted in the urine (humans) or faeces (cattle). Although the absorption of cobalt and iron involves a common transport pathway, it is possible that each element is also transported by specific transfer mechanisms.

Copper

Young mammals, with the exception of ruminants, have more copper in their tissues than adults. The human body contains 100 to 150 mg of copper. About two to four mg is taken each day in the diet, most of which is absorbed. Copper is absorbed mainly by the proximal small intestine in rats and dogs and by the distal small intestine in hamsters, and also by the stomach in dogs and hamsters.

The precise mechanism involving copper transfer is obscure. At low doses it is possibly absorbed by active transport and at high doses by simple (pore) diffusion. Intracellular transfer depends on metabolic energy. Copper is transported in the blood as a copper-protein (albumin) complex, stored in the liver and excreted mainly in the bile.

Iodine

The adult human body contains about 15 mg of iodine, over one-third of which is concentrated in the thyroid gland and the rest distributed in the tissues. About 125 μg of iodine are required each day. The salivary glands extract most of the iodine in arterial blood. Iodine is rapidly and completely absorbed, mainly as iodide, in the proximal small intestine. Iodine is required for the synthesis of the hormone thyroxine and other related compounds in the thyroid gland. Thyroxine is secreted from the liver in the bile. One-third is re-absorbed and the remainder excreted in the faeces and urine.

Manganese

The human body contains about 20 mg of manganese, found mainly in bone, liver and kidneys. The daily food intake yields five to ten mg of manganese, of which about one-third is rapidly absorbed, probably by active transport, and most of the rest excreted in the bile. Manganese and iron appear to have a common mechanism for absorption. The functions of manganese are uncertain but the ions are probably involved in oxidative phosphorylation.

Molybdenum

The human body contains about 15 mg of molybdenum. Molybdenum (and iron) forms part of the enzyme xanthine oxidase and is therefore required for the maintenance of normal levels of this enzyme in the intestine and liver (xanthine oxidase catalyzes the oxidation of nucleic acids). Molybdenum is easily absorbed in the small intestine by an unknown mechanism, and then rapidly transferred and stored in the liver, kidneys and bone. This element increases the absorption and excretion of phosphorus and decreases the store of copper in the liver.

Zinc

The human body contains about two g of zinc which is distributed in most tissues. The daily diet contains about 15 mg of zinc, which is absorbed mainly in the duodenum in monogastric animals and in both the duodenum and abomasum in digastric animals.

Little is known about the mechanism of zinc absorption and the results obtained are contradictory. Limited active transport or simple (pore) diffusion has been suggested. The amount of zinc absorbed varies widely, depending on a variety of factors such as species, age, and the amount of zinc reserves in the body. Zinc is rapidly lost over the entire length of the small intestine, possibly in the intestinal juice and in desquamated mucosal cells. Recent evidence using the everted rat jejunum indicates that zinc is secreted by an active process. Unabsorbed zinc is excreted mainly in the faeces.

17 Vitamins

Vitamins are essential, organic dietary constituents required in only very small amounts. They are important in intermediary metabolism and in the metabolism of certain tissues. Vitamins are either water-soluble (vitamin B complex, vitamin C) or fat-soluble (vitamins A, D, E and K). Little is known about their absorption and transport.

Water-soluble vitamins

These vitamins are readily absorbed, mainly into the portal blood. There appears to be no means of regulating their absorption.

Vitamin B complex

This group includes at least eleven different vitamins, some of which are taken in with the diet while others are synthesized by bacteria in the intestinal lumen. The most important vitamin, and about which most is known, is *vitamin B_{12}* (cobalamin). In man, about 15 μg per day are taken in by the food and up to five μg absorbed, which is slightly more than that required.

The absorption of vitamin B_{12} in adults is unique. In food, most of vitamin B_{12} is bound to protein, mainly as coenzyme B (cyanocobalamin) and must be released from peptide bonds before being absorbed. Because of their large molecular weight (M.W. 1350; molecular radius eight Å) the molecules cannot cross the water-filled pores (radius four Å) of the plasma membrane, and hence absorption by passive (pore) diffusion cannot occur. Most of the vitamin B_{12} molecules combine with a carrier glycoprotein, or *intrinsic factor* (IF), of even larger molecular weight (M.W. 55 000; molecular radius 36 Å), before being absorbed (Fig. 17.1). In man, cat, guinea-pig and rabbit IF is secreted by the oxyntic cells in the stomach under the influence of gastrin while in the mouse and rat it is secreted by the peptic cells. In unweaned rats, however, IF is not required for vitamin B_{12} absorption, and more of the vitamin is absorbed than in adults.

The mechanism of absorption takes place in three stages. (a) The separation of the vitamin from the protein in the food and its binding to IF in the stomach. Once formed the complex is less suceptible to enzyme action. (b) The rapid adsorption of the macromolecular B_{12}–IF complex onto specific receptor sites on or within the microvillous membrane of the distal ileum through a calcium and magnesium-dependent bond, a species-specific process independent of energy or enzyme-action. Insufficient IF leads to inadequate B_{12} absorption, and excess IF prevents absorption, for IF without the vitamin becomes attached to receptor sites, thus blocking the sites for the IF–B_{12} complex. (c) Intestinal transport, by an unknown but slow process, mainly in the distal ileum. The absorption of this vitamin may occur by pinocytosis, but the fact that the mechanism requires energy and is inhibited by lack of oxygen or low temperature suggests that an active process is likely. The presence of bile salts is necessary for absorption. To a much lesser extent absorption by passive (pore) diffusion appears to operate along the entire length of the intestine and is probably IF–independent.

The mechanism of transfer of the IF–B_{12} complex from the receptor sites, across the absorptive cell and into the bloodstream is largely unknown, but it is a

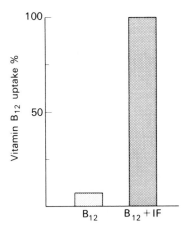

Fig. 17.1 Transport of vitamin B_{12} in the adult rat intestine either in the absence or presence of intrinsic factor (IF) (modified from Strauss and Wilson, 1960).

slow process and involves a complex series of events. IF probably separates from vitamin B_{12} during transit of the vitamin through the absorptive cell. This split takes place either at the apical membrane by the action of hydrolase enzymes or, more improbably, within the epithelial cell itself. In the portal blood the vitamin becomes bound to two different types of carrier-proteins (α- and β-globulin: transcobalamin I and II) and is rapidly transported in this form to the liver and stored there in large amounts. Some of the vitamin may be re-circulated in the bile and hence there is an enterohepatic circulation. The fate of IF after its release from the vitamin is uncertain; there is no evidence that it is absorbed.

Although the normal human dietary intake of *folic acid* (another vitamin B complex) is about 500 μg, only about 50 μg are required (an uncooked diet contains much larger doses of the vitamin). Folic acid is normally absorbed intact by the proximal small intestine, possibly by active transport. However, if taken in large doses it is absorbed passively along the entire intestinal length. Although the vitamin may undergo some metabolic changes during absorption it mainly enters the portal blood unaltered.

Other B-complex vitamins *(thiamine, riboflavin, pyridoxine* and *nicotinic acid)* and also vitamin C *(ascorbic acid)* are probably absorbed by simple (pore) diffusion in the proximal small intestine, although vitamin C might be transported by a carrier-mediated, sodium-dependent process. Vitamin C may be stored in the intestinal wall.

Fat-soluble vitamins

Vitamins normally found in association with dietary lipids include vitamins A, D, E and K. Bile salts are required for their transpo:t. Their absorption and passage into the circulation is poorly understood. They probably enter the absorptive cells incorporated as micelles. Within the cells they may become conjugated with proteins and enter the circulation as lipoproteins.

Vitamin A (retinol)

Vitamin A is derived from the hydrolysis of plant pigments known as β-carotenes (pro-vitamins), either in the intestinal epithelium (rodents, pigs and sheep) or in the liver (man). β-carotene molecules split into retinal molecules which undergo reduction to retinol and become converted to retinyl esters. These are absorbed into the lymphatic system by the proximal half of the small intestine, probably dissolved in chylomicrons formed during fat transport. Although vitamin A is assumed to be absorbed by simple diffusion through the lipid membrane, there is some evidence that an active transfer system is involved. About one-fifth of the β-carotenes are absorbed unchanged.

Vitamins D, E and K

These vitamins enter the intestinal epithelium unchanged, although a small proportion of vitamin D is esterified. Transport is probably by simple (lipid-soluble) diffusion in the proximal small intestine although recent work indicates that vitamin K, which is essential for maintaining normal blood coagulation, involves an energy-mediated process. Like vitamin A, the vitamins enter lymph channels but to a small extent all fat-soluble vitamins also enter the portal vein. Appreciable quantities of these vitamins remain unabsorbed and are excreted in the faeces. Man usually obtains enough vitamin D through ordinary exposure of the skin to sunlight. Vitamin D is associated with the transport of calcium and phosphorus.

18   Malabsorptive states

Many of the world's population have malabsorption at one time or another in their lives although it is often transient. Malabsorption occurs when the alimentary tract fails to function normally. This condition arises because the process of digestion is incomplete, or because it is impaired, or because both processes occur simultaneously. Failure to digest food arises following deficient pancreatic enzyme or bile salt secretion. A reduced flow of blood or lymph through the intestinal mucosa also affects absorption. Infectious animal diseases, particularly those caused by bacteria and helminth parasites, are common causes of malabsorption and other secondary effects. The resulting (but not invariable) damage to the mucosal absorptive surface and reduced digestive enzyme activity of the absorptive cells, frequently cause impaired absorption. Typically, the lamina propria of the villi becomes infiltrated by lymphocytes and plasma cells, and serum levels of immunoglobulins (IgA) are elevated.

Manifestations of malabsorption include loss of appetite and food intake, diarrhoea, steatorrhoea, abdominal pains, vomiting and weight loss. Significant losses of fluid in the small intestine lead to severe dehydration. Unfortunately, current malabsorption tests are often inadequate. In man, a faecal fat loss greater than seven g in 24 hours in the faeces is a diagnostic feature of steatorrhoea. A flat blood sugar curve after taking an oral dose of D-glucose, or reduced D-xylose excretion in the urine following an oral dose of the sugar, also give indications of impaired absorption. Other diagnostic symptoms include lowered serum iron and vitamin B_{12}. Patients with suspected malabsorption are often given a diluted barium mixture and the transit of the barium preparation is traced by taking a radiographic film of the bowel at regular intervals. Routine conventional microscopy or enzyme analysis, particularly that of disaccharidase activity of small pieces of intestinal homogenates, are occasionally made to diagnose malabsorptive conditions. Standard *in vivo* and *in vitro* methods for assessing normal intestinal absorption (Chapter 11) are also frequently used for investigating experimental malabsorption in diseased animals.

Primary malabsorption

Primary malabsorption manifests itself in temperature regions under a variety of syndromes, collectively referred to as coeliac sprue, and in tropical regions as tropical sprue. Infants and certain adult races may also lack one or more intestinal enzymes that hydrolyze certain disaccharides, a condition termed disaccharide intolerance.

Coeliac sprue (coeliac disease, adult coeliac disease, idiopathic sprue, ideopathic steatorrhoea, nontropical sprue, gluten-induced enteropathy)

This is a congenital disease, possibly of bacterial or viral origin, limited to temperate regions, in which children and adults become sensitive to gluten (a toxic factor present in gliadin, a component of gluten) in ordinary flour. Symptoms include diarrhoea and steatorrhoea, weakness and weight loss. Children become potbellied and the buttocks become wasted. Young adult white female races are more susceptible to the disease than others.

The incompletely digested gluten damages the lining of the proximal small intestine by flattening the mucosal surface. There is a ten-fold difference between the area of mucosal surface of normal patients and those with coeliac sprue. The villi shorten, the epithelial cells become cuboid, the microvilli are disrupted and the lamina propria is oedamatous and infiltrated by leucocytes. Although the absorptive cells are formed more quickly in the crypts they migrate more slowly along the villus and so persist longer than normal cells; hence they are aged. Adults show malabsorption of D-xylose, amino acids, fats, folic acid and vitamin A. Intestinal disaccharidases become reduced and there is a net secretion of fluid and electrolytes into the intestinal lumen. Removing gluten from the diet gives a dramatic clinical response.

Tropical sprue

A malabsorption syndrome occuring particularly in white-skinned adult males living between latitudes 30°N and 20°S in the Caribbean, southern India and south-east Asia, but not usually Africa. Patients show symptoms of chronic diarrhoea, steatorrhoea, a sore tongue, lassitude and weight loss. These symptoms are associated with impaired absorption of carbohydrates, fats, folic acid and vitamin B_{12}. The origin of the disease is uncertain but it is probably bacterial since there is a quick response to antibiotic treatment. The proliferation of bacteria in the lumen of the small intestine, that originate from the large intestine, suggests that the disease is associated with an abnormal small intestinal bacterial population. Both environmental and nutritional conditions may also become associated with the disease. The intestinal mucosa sometimes becomes flattened and intestinal disaccharidases are reduced. Patients respond to antibiotic, folic acid and vitamin B_{12} therapy. There is little response to a gluten-free diet.

Disaccharide intolerance

This disease arises as a failure of the small intestinal surface to hydrolyze certain disaccharides in the diet because one or more specific enzymes are missing in the apical membrane of the absorptive cells. The histology of the mucosa remains unchanged. Disaccharide malabsorption is either inherited (congenital) or acquired. In infants the disease is the result of an inherited mucosal abnormality; in adults the disease is acquired and usually transitory. Secondary disaccharide deficiency may also occur in other diseases.

Deficiency of a specific enzyme results in the corresponding disaccharide remaining partly undigested, causing severe osmotic diarrhoea and consequent malnutrition and dehydration. The undigested sugar passes into the distal ileum and large intestine to be partly hydrolyzed by bacterial enzymes and undergo bacterial fermentation into organic (mainly acetic and lactic) acids. Both osmotic and fermentative diarrhoea leave the faeces liquid and acid. Two main disaccharidase deficiencies are known:

(a) LACTASE DEFICIENCY Congenital lactose intolerance is rare in infants but common after weaning. There are distinct racial differences in the prevalence of lactase deficiency in adults. The condition is common in non-white populations of the world who are unable to tolerate large quantities of milk. There is uncertainty over the etiology of the defect; it may be either genetically determined or an acquired deficiency owing to the absence of lactose at a critical stage during development. Lactase deficiency does not appear to be adaptive in humans but can be induced in rats.

(b) SUCRASE-ISOMALTASE DEFICIENCY This rare inherited deficiency in humans is present at birth but symptoms do not appear until sucrose is included in the diet. Although isomaltose is not found in food symptoms appear with a carbohydrate-rich diet since isomaltose is released from branching points of starch during its digestion by α-amylase. It is not an acquired deficiency but possibly an inborn error of metabolism.

Secondary malabsorption

Secondary malabsorption occurs in a wide variety of pathological conditions, during which deterioration to the normal small intestinal mucosa results in symptoms of impaired absorption. Damage to the mucosa is usually localized but sometimes widespread, giving rise to ulceration, inflammation, and cellular infiltration or fibrosis. The affected villi are frequently partially or totally atrophied, thus flattening the mucosal surface, and the lamina propria becomes oedamatous and inflamed. Occasionally, malabsorption is evident despite a normal mucosa.

Bacteria, fungi, viruses, protozoa, helminth parasites, mechanical defects of the small intestine, some organic disorders, drugs, antibiotics and other chemicals are all capable of producing manifestations of malabsorption. Secondary malabsorptive states may also originate from non-enteric conditions that affect the stomach, pancreas, liver and gall bladder. The effects of mainly bacteria, protozoa and helminth parasites on intestinal absorption will be briefly considered.

Normal bacteria of the alimentary canal and their functions

The lumen of the human alimentary canal contains a varied population of about 60 species of non-pathogenic bacteria, which are commensal or symbiotic parasites. Since many of them live within the mucous layer of the intestinal surface, sampling the luminal fluid will not give a true estimate of their numbers. The empty stomach is virtually free from them (in man $< 10^3$ per cm^3) since bacteria that are swallowed are largely destroyed by the corrosive action of the gastric juice. The duodenum and jejunum have a sparse, transient bacterial population ($< 10^5$ per cm^3) consisting mainly of Gram-positive aerobic bacteria that are being continually propelled towards the ileum and destroyed by the combined action of the gastric juice and bile acting as disinfectants. The ileum represents a transitional region between the sparse bacterial population of the proximal small intestine and the luxuriant bacterial population of the large intestine. In the ileum there is a relatively large, stable population (10^6 to 10^8 per cm^3) of approximately equal numbers of aerobic and anaerobic bacteria. In man, the large intestine and caecum contain a large, resident population (10^9 per cm^3) of mainly Gram-negative anaerobic bacteria, whereas the large intestine and caecum of other mammals have a mixed bacterial population, like the ileum, owing to coprophagy, compartmentalized stomachs, and environmental differences in the lumen. Finally, wet faeces contain a heavy, mixed, bacterial population ($> 10^{10}$ per g) amounting to one-third of the faecal dry matter. There are wide variations in the normal bacterial population of healthy indivic..als in the tropics, depending on the nutritional status of the host.

Bacteria can alter the absorption of substances in the normal small intestine in several ways. Firstly, they may affect the physical milieu within the intestinal lumen. Thus, changes in pH caused by bacterial metabolism can alter the ionization of food molecules. Secondly, they may affect the development of the epithelial cells of the villus. For example, the mucosal surface area in germfree rats is significantly less than in conventional animals owing to the reduced rate of cell proliferation. Thirdly, the metabolic products of the bacteria may exert local reactions in the intestine or become absorbed and act as catalysts to certain enzyme systems of the host. Intestinal bacteria are also responsible for altering some substances in the lumen which may become utilized by the host. Thus, bacteria hydrolyze conjugated bile acids, and convert bilirubin to urobilinogen. They synthesize several vitamins, including vitamins B_{12}, K, and some folic acid. Finally, intestinal bacteria can also stabilize secretory immunoglobulins of the host.

Bacterial diseases of the small intestine

Bacterial infection is the most widespread of all alimentary tract diseases, particularly among the young. The pathogen is usually taken in contaminated food or drinking water. Although many bacteria are normally destroyed by the acid-content of the stomach, bile acids and secretory immunoglobulin A, some will nevertheless reach the small intestine. An incubation period follows varying from less than a day to over a week during which the organisms proliferate rapidly in the intestinal lumen and liberate enterotoxins consisting of large protein-polypeptide molecules (M.W. 30 000–100 000). The intestinal mucosa often becomes eroded or ulcerated, either locally or extensively, and the tissues become oedomatous and infiltrated. The enterotoxins poison the individual by entering the bloodstream through the damaged mucosal surface. Bacteria themselves frequently penetrate the affected area and also pass into the circulation. The irritation caused by the invading organisms increases intestinal motility, which serves to expel the intestinal contents at frequent intervals.

The onset of the infection causes fever, abdominal pains, vomiting and eventually dehydration, malabsorption and weight loss. A dysenteric attack begins with diarrhoea, which may be acute, intermittent or chronic, during which the individual passes frequent loose stools which are watery and contain mucus and blood. Depending on the pathogen, the illness may be short-lived or prolonged, mild or severe. Diarrhoea in man is usually treated by a sojourn in bed without solid food for one to two days, sipping water and taking antibiotics.

Many bacterial diseases cause impaired balance between absorption and secretion. Malabsorption of solutes is occasionally accompanied by net secretion of fluid and electrolytes. Impaired absorption and excess fluid secretion are sometimes located at different sites along the small intestine. Exceptionally, in cholera the mucosa remains intact, secretion of fluid is profuse but absorption of solutes seems unimpaired. The undifferentiated cells in the crypts of Lieberkühn are possibly responsible for excess fluid secretion in some bacterial diseases. The mechanism causing fluid secretion is unclear although hypersecretion , initiated by enterotoxins, is thought to be stimulated by the increased activity of adenyl cyclase, an enzyme located in the lateral plasma membrane of the absorptive cells, which converts ATP to cyclic AMP.

Some of the symptoms in patients affected by intestinal bacterial infections have been mimicked in experiments on laboratory animals. Starvation, alkali agents, opiate or antibiotic pretreatment have been used singly or in combination to counteract the natural resistance of the animals to challenge by bacterial infection.

CHOLERA Cholera is a waterborne disease caused by *Vibrio cholerae* bacteria which secrete the toxin choleragen. The disease is prevalent in eastern countries, when dehydration is accompanied by voiding profuse quantities of fluid in 'rice water' stools. Studies in man and experimental animals have shown that, although the mucosa of the infected small intestine remains normal*, there is a pronounced net secretion of water and electrolytes into the intestinal lumen, but the absorption of water, sugars and amino acids apparently remains unaltered, as

* The transmission electron microscope has recently shown that this disease produces small pseudopod-like cytoplasmic processes projecting from the apical surface of the absorptive cells into the lumen. Other cellular defects were also found.

does the electrochemical potential gradient, suggesting that hypersecretion rather than underabsorption is the cause of diarrhoea.

Oversecretion in cholera possibly occurs from the undifferentiated cells in the crypts of Lieberkühn in response to stimulation of the enzyme adenyl cyclase by the cholera toxin that converts ATP to cyclic AMP. The phenomenon of unimpaired glucose absorption with net fluid secretion in cholera has proved to have therapeutic value. Feeding glucose-rich Coca Cola syrup to patients alleviates dehydration because normal glucose absorption enhances sodium transport with water absorption following isomotically.

SALMONELLOSIS (FOOD POISONING AND ENTERITIS) The Salmonella group, of which about 1 500 species have been isolated, occurs naturally in the small intestine of many mammals, although the incidence of infection is low. *Salmonella typhimurium* is a common cause of food-poisoning from infected prepared foods, particularly meat and egg products. The bacteria multiply in the infected food when it is left at a warm temperature. The development of mass catering has increased the spread of the disease.

The organisms proliferate in the small intestine, disrupt and penetrate the mucosa, invade the bloodstream and cause fever, vomiting, a watery diarrhoea, dehydration and malabsorption. Experiments in mice infected with *S. enteritidis* have shown that the whole length of the small intestinal mucosal surface became eroded and disrupted (Fig. 18.1). The absorption of different hexoses and amino acids became markedly impaired. Malabsorption of D-glucose occurred throughout the everted small intestinal length; although fluid transfer decreased in the proximal and mid-intestine, it increased distally (Fig. 18.2). This did not occur in the experimental controls. Different disaccharidases also became markedly reduced; disaccharide absorption was impaired and the electrical potential gradients decreased.

Widespread diarrhoeal diseases in farm livestock are frequently caused by different groups of *Salmonella* and may lead to significant economic losses. Calves and sometimes adult cattle become infected during the summer months with a chronic, infective, haemorrhagic enteritis caused mainly by *S. dublin* (calf paratyphoid). Pigs, particularly after weaning, similarly become affected by *S. cholera-suis* (syn. *suipestifer*) which invade the mucosa and penetrate the lymph spaces (swine typhoid, pig typhus). The incidence of infection in sheep is low. Presumably these infections also cause impaired absorption.

STAPHYLOCOCCAL FOOD-POISONING AND ENTERITIS *Staphylococcus aureus* occasionally contaminates commercially prepared foodstuffs, particularly in the U.S.A., and causes a short but severe illness. Following infection, prostration, violent abdominal pains, vomiting and sometimes diarrhoea quickly follow. The condition is caused by heat-stable enterotoxins (mainly enterotoxin A) liberated by the bacteria in the food. The organism does not usually affect the small intestinal structure or function, although excess fluid may be secreted. However, everted intestinal sacs of normal rats incubated with staphylococcal toxin

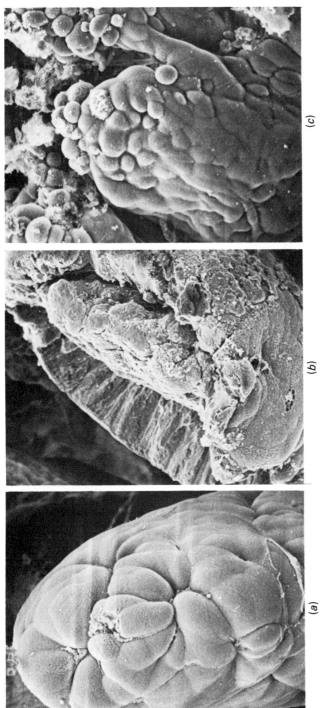

(a)

(b)

(c)

Fig. 18.1   (a) Normal villous tip of mouse small intestine. The surface is thrown into interconnecting ridges (sulci). The orifices are probably the openings of the goblet cells. The slight disruption at the tip probably represents normal shedding of epithelial cells. Approximately × 1 360.

(b) Villous tip of proximal small intestine of mouse infected with salmonellosis. The tip is ulcerated, exposing the sides of the epithelial cells and the central core of the lamina propria. Approximately × 130.

(c) Villous tip of distal small intestine of mouse infected with salmonellosis. The tip is disrupted and assumes a beaded or frothy-like appearance. Approximately × 1 050. Scanning electron microscopy. (Reproduced by permission from Madge, D.S. (1974) *J. Microscopie* **20**, 45—50).

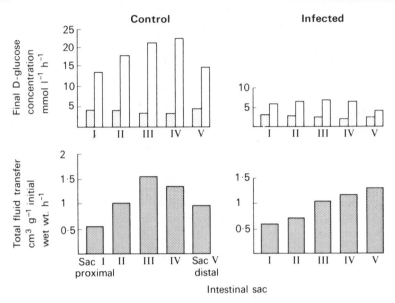

Fig. 18.2 Effect of *Salmonella enteritidis* infection (5 × 10^9 each) on everted *in vitro* small intestinal uptake of D-glucose and total fluid (glucose dependent+glucose independent) transfer in adult mice. *Above* Each paired histogram represents the final mucosal (left) and final serosal (right) concentration of glucose after 1 h at 37°C. Initial concentration of glucose: 8.0 mmol l^{-1}. The small intestine was divided into five sacs of equal length. Sac I: proximal: sac V: distal. Average values for 15 mice. *Below* Total fluid transfer. Average results expressed in cm^3 g^{-1} initial wet weight h^{-1}. Average values for 15 mice (modified from Madge, 1972).

(enterotoxin B) showed reduced absorption of glucose, water and electrolytes. In contrast, hexose and amino acid absorption remained unaltered in mice following inoculation with prototype staphylococci producing enterotoxins A, B, C or D. A group (Group III) of staphylococci can sometimes cause a fatal gastroenteritis following injudicious use of broad-spectrum antibiotics that allow intestinal overgrowth of the organism.

CLOSTRIDIAL FOOD-POISONING AND ENTERITIS Type A strains of *Clostridium perfringens* (syn. *welchii*) have achieved notoriety as an important cause of food-poisoning. The organism, mainly found in cooked meat and poultry contaminated with soil or faecal matter, releases a potent toxin. Both heat-sensitive and heat-resistant strains cause diarrhoea. Fortunately, the illness is mild and soon over. Vomiting is unusual. Aside from net fluid secretion, the small intestine remains virtually unaltered.

Types B, C and D strains colonize the intestine of sheep following sudden changes in the diet and cause enterotoxaemia. This condition leads to increased permeability of the intestinal mucosa, during which toxins are absorbed. The mucosa usually remains unchanged, but may sometimes become ulcerated and haemorrhagic in lambs, causing lamb dysentery. Type C strain can also cause haemorrhagic enteritis in piglets and calves.

SHIGELLOSIS (BACILLARY DYSENTERY) The group *Shigella*, particularly *Shigella sonnei*, is responsible for the commonest mild form of dysentery in man in England and Wales, over 20 000 cases being notified each year for at least two decades. The disease is prevalent among children and transmitted by faecal-oral contamination. Symptoms include fever, abdominal cramp and vomiting, followed by dysentery.

Although difficult to detect in the intestinal tissues, the bacteria produce enterotoxins that enter the circulation mainly through micro-ulcers on the mucosal surface. Like many other bacillary infections, fluid and electrolyte secretion is accompanied by reduced absorption of food substances. Mice experimentally infected with *S. dysenteriae* or *S. flexneri* developed transient malabsorption in one to two weeks after which time the animals recovered their ability to absorb normally. Like those affected by *Salmonella enteritidis* they showed uniformly depressed *in vitro* D-glucose absorption, and fluid transfer in everted sacs fell proximally but rose distally.

INFANTILE GASTROENTERITIS *Escherichia coli* is normally found in the alimentary canal of all mammals. On the basis of their antigenic properties the organism can be divided into a large number of serogroups and serotypes. Only a few strains are asociated with gastroenteritis. It is unclear why certain strains are enteropathogenic while others are harmless. Sporadic outbreaks of the disease occur during summer and autumn in infants living in sub-standard hygienic conditions; outbreaks in adults are rare.

Rapid proliferation of *E. coli* in both the small and large intestines causes a severe, diffuse enteritis. The main symptoms are an acute, undifferentiated diarrhoea and vomiting, with resulting dehydration and weight loss. The mucosal surface of the small intestine sometimes remains unchanged although the organism may become plastered over the surface of the villi and proliferate on or within the overlying mucous layer. Sometimes, however, the infection results in inflammatory lesions, and causes a fatal enteritis with severe diarrhoea, as, for example, in newly born calves (white scours). Virulent organisms elaborate a toxin that produces fluid accumulation in ligated loops of the small intestine of rats, rabbits, pigs, calves and lambs. There is net secretion of fluid and electrolytes, and glucose and vitamin B_{12} absorption is impaired.

TUBERCULOSIS AND PARATUBERCULOSIS Acute intestinal tuberculosis is caused by *Mycobacterium tuberculosis* which invade the mucosa in the ileocaecal region and form oval-shaped ulcers. Infected tissues become destroyed and replaced by aggregations of fibrous tissue and lymphocytes. Malabsorption of fat and vitamin B_{12} is sometimes evident.

Johne's disease (paratuberculosis) is a chronic, infective enteritis in ruminants, particularly cattle, leading to significant economic losses of livestock. Infected animals suffer from emaciation and persistent diarrhoea. The disorder is caused by invasion of *Mycobacterium johnei* in the small intestinal mucosa and associated lymph glands, resulting in a thickened, haemorrhagic wall and a corrugated inner surface. These structural changes are accompanied by severe malabsorption of amino acids and hence the animals become thin and wasted. Mice infected with the disease showed decreased hexose and amino acid absorption and reduced fluid transfer. The intestinal epithelium became eroded in patches and the mucosa was slightly inflamed. Clumps of bacteria were seen in the lumina propria and submucosa.

TYPHOID AND PARATYPHOID *Salmonella typhi* is a pathogen limited to man and usually transmitted in drinking water or milk. The bacteria pass from the small intestinal mucosa into Peyer's patches, enter the circulation where they proliferate, reach the liver, and are voided from the bile into the intestinal lumen. A secondary invasion of the Peyer's patches follows during which they become swollen and ulcerated. Fever in patients is remittent and they void diarrhoetic 'pea green' stools.

Paratyphoid is caused by *Salmonella paratyphi*. The disease is less severe although similar to typhoid.

WHIPPLE'S DISEASE This condition affects mainly middle-aged white males. Symptoms include dehydration, fever, and weight loss. Although the disease is a generalized one, the alimentary tract is mainly affected. The small intestine becomes dilated and the wall thickened. The villi in the proximal region may show partial atrophy, but the absorptive cells appear normal. Characteristically, the lamina propria becomes filled with large, pale, foamy cells (macrophages) which appear granular when stained with the periodic acid-Schiff reagent. Minute, rod-shaped bacillary bodies are also evident in the lamina propria, and the lacteals are distended.

CROHNE'S DISEASE Patients with this condition suffer from lethargy, abdominal pains, vomiting, diarrhoea and weight loss, accompanied by malabsorption of fat and fat-soluble vitamins. Any region of the alimentary tract may become deranged, and the ileums is often affected. 'Hose pipe' strictures develop that narrow the lumen of the ileum; the mucosal surface becomes ulcerated and the submucosa contains aggregations of lymphocytes in the capillaries. The origin of the disease is unknown, although a variety of stimuli, including pathogenic bacteria, have been suggested.

Blind (stagnant) loop syndrome

Certain conditions either interfere with orderly intestinal peristalsis, or form a loop or diverticulum of the small intestine, within which massive proliferation (10^{11} per cm^3) of different kinds of bacteria is manifest. This condition is associated with vitamin B_{12} malabsorption which causes anaemia and other defects, including diarrhoea and impaired carbohydrate and fat absorption.

Although malabsorption can be corrected by antibiotics or surgical treatment, the precise mechanism by which bacteria cause impaired absorption is unknown. However, they may produce toxic metabolites that interfere with intestinal epithelial function, or the bacteria may compete for the vitamins in the intestinal lumen. Substantial quantities of nutrients are required to support bacterial growth and multiplication and so the host is deprived of sources of nutrition. The mucosal surface is unaltered and the villi are apparently normal but the lamina propria becomes invaded by leucocytes.

Protozoa

GIARDIASIS This disease is caused by invasion of a flagellate protozoan, *Giardia lamblia* (syn. *intestinalis*), in the proximal small intestine. The infection is more common in children than in adults, particularly in unhygienic conditions. Symptoms include recurring diarrhoea, steatorrhoea, abdominal distension and mild malabsorption of D-xylose, glucose and vitamin B_{12}. The condition may produce a flat mucosa, local inflammation, and lowered disaccharidases. The villi sometimes become short and irregular and the microvilli are disrupted. The parasite attaches itself to the mucosal surface by a ventral sucker and feeds through its dorsal surface. The parasites mainly affect absorption by providing a mechanical barrier over the mucosal surface and by nutritional competition between host and parasite.

MALARIA Patients with malignant malaria (*Plasmodium falciparum*) frequently have diarrhoea, suggesting a malfunction of the small intestine. Vasoconstriction of the intestinal blood vessels is thought to be partly responsible for this condition. Mice and monkeys with malaria show impaired absorption of D-xylose, folic acid, and vitamin B_{12}; inflammatory cells are seen in the lamina propria.

Helminth infections of the small intestine

Many helminth (worm) parasites can infect the mammalian small intestine. Heavy infections produce loss of appetite, lassitude, abdominal pains, vomiting, diarrhoea and weight loss. Medical and veterinary evidence indicate that the pathology of helminthic parasitism is poorly understood.

Although helminths usually cause damage to the absorptive intestinal surface, few experiments on intestinal malfunction, particularly with farm livestock, have been attempted. Hence, only the known effects of certain helminth parasites on intestinal absorption are given, although the intestinal damage produced by many other helminths suggests that malabsorption of nutrients is widespread.

The effects of helminth infections on intestinal absorption are sometimes both conflicting and confusing as they often cause variable responses between different individuals. In some, absorption is markedly reduced; in others, absorption is normal. Reproducible results are often difficult to obtain owing to the patchy distribution of the damage on the mucosal surface and because malabsorption is dose-dependent. Impaired uptake of solutes is usually accompanied by depressed fluid absorption and not excess fluid secretion, as in some bacterial infections. The diagnosis of gastrointestinal helminthiasis depends on the presence of characteristic eggs in the faeces or in the recovery of adult worms.

Trematoda

SCHISTOSOMIASIS An individual becomes infected while wading or bathing in water contaminated with the larval stages of the worm. The larvae penetrate the skin, enter the bloodstream and reach the liver where they mature into adults. They then migrate into the mesenteric veins of the small and large intestines and lay eggs that pass through the intestinal wall into the faeces. The eggs cause a local inflammatory reaction with surrounding fibrotic lesions. The villi, however, remain unchanged. Symptoms include fever, fatigue and a bloody diarrhoea; D-xylose and fat malabsorption are uncommon. In mice affected by *Schistosoma mansoni* the small intestinal wall became thicker but the villi were unchanged, and absorption of glucose, methionine and sodium propionate remained unimpaired.

Nematoda

ANKYLOSTOMIASIS (HOOKWORM DISEASE) Hookworm infections is an important cause of ill-health in the tropics and sub-tropics. Adult worms bite into the lining of the proximal small intestinal wall in man and ruminants, release an anti-coagulant, suck the blood of the host and cause intestinal bleeding. Fertilized eggs are voided in the faeces; larvae hatch and develop in the soil to eventually enter the host through the skin or by mouth.

Owing to loss of blood the disease causes iron deficiency anaemia. The small intestinal mucosa occasionally becomes eroded, the villi become leaf-shaped and the lamina propria inflamed. D-xylose, fat, and vitamin A absorption are sometimes impaired. Dogs infected with *Ankylostoma caninum* developed small ulcerations, bleeding spots and blunting of the villi, mainly in the proximal small intestine, defective absorption of carbohydrates, amino acids and sometimes fats, and excess mucus production. Malabsorption was dose-dependent.

TRICHINOSIS (TRICHINIASIS) The adult worms *(Trichinella spiralis)* live in the proximal small intestine in man and other mammals. The female burrows into the intestinal wall, liberates young larvae into the blood-lymph circulation, which are distributed all over the body to eventually encyst in voluntary muscle. The disease causes weight loss, diarrhoea and other malfunctions. Mice and guinea-pigs infected with the parasite showed impaired absorption of glucose; malabsorption was dose-dependent. The villi became shorter and wider and the absorptive cells were flattened although they remained intact.

STRONGYLOIDIASIS A fairly common disease in the tropics and sub-tropics, when heavy infections can cause death by toxaemia and septicaemia. Symptoms include diarrhoea alternating with constipation, vomiting and weight loss. The nematode, *Strongyloides stercoralis,* enters the body by the skin, passes through the lungs, up the trachea and is swallowed, to reach the proximal small intestine where the females mature, lying deeply buried in the mucosa. Eggs hatch in the intestinal lumen and, in humans, the young larvae may invade the mucosa, causing extensive lesions, abnormal villi, oedema and inflammation. The disease can cause malabsorption of D-xylose and folic acid.

NIPPOSTRONGYLOSIS *Nippostrongylus brasiliensis* (syn. *muris*) is a nematode parasite of rats. Larvae, from eggs in the faeces, develop into a free-living infective stage that enter the host in the skin, pass into the bloodstream and migrate through the lungs on their way to the proximal small intestine. L. E. A. Symons (1969) has studied in detail the effects of the infection in rats. Diseased rats lost their appetite and body weight and became diarrhoetic. The small intestine became congested and filled with fluid. The rate of passage of food through the jejunum in infected rats was slower than that through the worm-free ileum, which was faster than in the controls. The mucosa of the proximal small intestine became flattened in patches, the villi were stunted and the underlying tissues became oedamatous and infiltrated. The distal small intestine appeared normal.

Over the entire small intestine there was no significant decrease in glucose or histidine absorption, although the animals were severely diseased. Pancreatic enzymes (amylase and trypsin) were unchanged, but intestinal disaccharidase and peptidase levels became reduced, and digestion of maltose and protein decreased. However, in other experiments the jejunum showed depressed net uptake of glucose, water and electrolytes, whereas in the ileum the rate of glucose, water and electrolyte absorption increased. Thus, a compensatory mechanism in the unparasitized distal end corrected the faults in the parasitized proximal end, so that the overall small intestinal function remained unaltered.

Protein-losing enteropathy

Normal protein losses from the small intestine, collectively referred to as protein-losing enteropathy, are significantly increased in at least 70 different gastrointestinal diseases, including many bacterial and helminth infections. Ulceration and other mucosal surface changes cause excessive loss of plasma proteins into the small intestinal lumen which are rapidly degraded to their constituent amino acids, to be partly reabsorbed and made available to the body for protein synthesis. Protein deficiency occurs when the rate of protein loss exceeds that of protein synthesis. Proteins may also become depleted in malfunctions not mainly affecting the gastrointestinal tract.

Das wenige verschwindet leicht dem Blicke
Der vorwärts sieht, wie viel noch übrig bleibt. (G. W. Goethe)

The little that is done seems nothing
When we look forward, and see how much we have yet to do. (Trans. M. Arnold)

POSTSCRIPT

A recently published report by the joint Agricultural Research Council and Medical Research Council committees on *Food and Nutrition Research* (HMSO, London, 1974) is both timely and apposite. The report outlines the multidisciplinary approach to basic research on gastrointestinal functions, defines practical problems in human and animal nutrition, and summarizes outstanding priorities and recommendations for future work.

Further reading

Section 1: Secretion, Digestion and Motility

GENERAL

BOGOCH, A. ed. (1973). *Gastroenterology*, McGraw-Hill Book Co., New York.

DAVENPORT, H. W. (1966). *Physiology of the Digestive Tract*. 2nd edition. Year Book Medical Pub. Inc., Chicago.

GREGORY, R. A. (1962). *Secretory Mechanisms of the Gastro-intestinal Tract*. Edward Arnold (Pub.) Ltd., London.

PASSMORE, R. and ROBSON, J. S. eds. (1968). *A Companion to Medical Studies*. Volume 1. Chapter 30. Blackwell Scientific Publications, Oxford.

PHILLIPSON, A. T. ed. (1970). *Physiology of Digestion and Metabolism in the Ruminant*. Oriel Press, Newcastle upon Tyne.

PIKE, R. L. and BROWN, M. L. (1967). *Nutrition: An Integrated Approach*. John Wiley & Sons, Inc., London.

SPENCER, R. P. (1960). *The Intestinal Tract*. C. C. Thomas Publications, Springfield, Illinois.

SUMMERSKILL, W. H. J. Course director (1973). Gastroenterology: Digestion and Disorders of Digestion in Man. *Mayo Clin. Proc.* **48,** 589–662.

SWENSON, M. J. ed. (1970). *Dukes's Physiology of Domestic Animals*. Part III. Comstock Publishing Associates, Cornell University Press, Ithaca, Cornell.

TEXTER, E. JR., CHOU, C. C., LAURETA, H. C. and VANTRAPPEN, G. R. (1968). *Physiology of the Gastrointestinal Tract*. C. V. Mosby Co., St. Louis, Minnesota.

TRUELOVE, B. C. and JEWELL, D. P. eds. (1973). *Topics in Gastroenterology 1*. Blackwell Scientific Publications, Oxford.

A detailed bibliography is not given, but useful reviews are found in the *Annual Review of Physiology*. These include:

SECRETION

ANDERSSON, S. (1973). Secretion of gastrointestinal hormones. *A. Rev. Physiol.* **35,** 431–52.

FARRAR, E. and BLOWER, R. J. (1967). Gastric juice and secretion. *A. Rev. Physiol.* **29,** 141–68.

GREGORY, R. A. (1965). Secretory mechanisms of the digestive tract. *A. Rev. Physiol.* **27,** 395–414.

HENDRIX, T. R. and BAYLESS, T. M. (1970). Digestion: intestinal secretion. *A. Rev. Physiol.* **32,** 139–64.

MOTILITY

BORTOFF, A. (1972). Digestion: motility. *A. Rev. Physiol.* **34,** 261–90.
DANIEL, E. S. (1969). Digestion: motor functions. *A. Rev. Physiol.* **31,** 203–26.
MENGUY, R. (1964). Motor functions of the alimentary tract. *A. Rev. Physiol.* **26,** 227–48.

MUSCLE AND NERVES

BURNSTOCK, G. and HOLMAN, T. S. (1963). Smooth muscle: autonomic nerve transmission. *A. Rev. Physiol.* **25,** 61–90.
PROSSER, C. L. (1974). Smooth muscle. *A. Rev. Physiol.* **36,** 503–35.

Section 2: Absorption

MONOGRAPHS

BROOKS, F. P. (1970). *Control of Gastrointestinal Function.* MacMillan Co., London.
GLASS, G. B. J. (1968). *Introduction to Gastrointestinal Physiology.* Prentice–Hall Inc., New Jersey.
WILSON, T. H. (1962). *Intestinal Absorption.* W. B. Saunders & Co., Philadelphia.
WISEMAN, G. (1964). *Absorption from the Intestine.* Academic Press, London.

SYMPOSIA

BURLAND, W. L. and SAMUEL, P. D. eds. (1972). *Transport Across the Intestine.* Churchill Livingstone, London.
DAWSON, A. M. ed. (1971). Intestinal Absorption and its Derangements. *J. clin. Path.* **24,** Suppl. (Roy. Coll. Path.) 5.
McD. ARMSTRONG, W. and NUNN, A. N. JR. (1971). *Intestinal Transport of Electrolytes, Amino Acids and Sugars.* C. C. Thomas Publications, Springfield, Illinois.
SMYTH, D. H. ed. (1974). *Intestinal Absorption* (2 vols.). Plenum Publications Co. Ltd., London.
SMYTH, D. H. ed. (1967). Intestinal Absorption. *Br. med. Bull. (Brit.)* **23** (3), September 1967.

Below is a list of original papers and monographs which represent only a small proportion of the total consulted. They supplement the works given above.

VILLOUS STRUCTURE

CHENG, H. and LEBLOND, C. P. (1974). Origin, differentiation and renewal of the four main epithelial cells types in the mouse small intestine. *Am. J. Anat.* **141,** 461–561 (five papers).
CROFT, D. N. and COTTON, P. B. (1973). Gastro-intestinal cell loss in man. *Digestion* **8,** 144–60.
FORSTNER, G. L. (1969). Surface sugars in the intestine. *Am. J. med. Sci.* **258,** 172–80.

ISOMÄKI, A. M. (1973). A new cell type (tuft cell) in the gastrointestinal mucosa of the rat. *Acta path. microbiol. scand.* Sect. A, suppl. No. 240, 1–35.

ITO, S. (1969). Structure and function of the glycocalyx. *Fedn Proc. Fedn. Am. Socs exp. Biol.* **228,** 12–25.

JOHNSON, C. F. (1967). Hamster intestinal brush-border surface particles and their function. *Fedn Proc. Fedn. Am. Socs exp. Biol.* **28,** 26–29.

LIPKIN, M. (1965). Cell replication in the gastrointestinal tract of man. *Gastroenterology* **48,** 616–24.

PEARSE, A. G. E., COULLING, I., WEAVERS, B. and FRIESEN, S. (1970). The endocrine polypeptide cells of the human stomach, duodenum and jejunum. *Gut,* **11,** 649–58.

TONER, P. J. and CARR, K. E. (1971). *Cell Structure.* 2nd edition. Churchill Livingstone, London.

TONER, P. G. (1968). The cytology of intestinal epithelial cells. *Int. Rev. Cytol.* **24,** 233–343.

TRIER, J. S. (1968). Morphology of the epithelium of the small intestine. In *Handbook of Physiology.* Ed. C. F. Code. Section 6, Vol. 3, Chapter 63. Pub. Am. Physiol. Soc., Washington, D.C.

PLASMA MEMBRANE AND TRANSPORT

BARRY, R. J. C., SMYTH, D. H. and WRIGHT, E. M. (1965). Short-circuit current and solute transfer by rat jejunum. *J. Physiol., Lond.* **181,** 410–31.

BRETSCHER, M. S. (1973). Membrane structure: some general principles. *Science* **181,** 622–29.

CAPALDI, R. A. (1974). A dynamic model of cell membranes. *Scient. Am.* **230** (March 1974), 27–33.

CHARNEY, A. N., GOTS, R. E. and GIANNELLA, R. A. (1974). $(Na^+\!\!-\!\!K^+)$—stimulated adenosinetriphosphatase in isolated intestinal villus tip and crypt cells. *Biochim. biophys. Acta* **367,** 265–70.

CRANE, R. K. (1962). Hypothesis for mechanism of intestinal active transport of sugars. *Fedn Proc. Fedn. Am. Socs exp. Biol.* **21,** 891–95.

CRANE, R. K. (1965). Sodium-dependent transport in the intestine and other animal tissues. *Fedn Proc. Fedn. Am. Socs exp. Biol.* **24,** 1000–6.

CSAKY, T. Z. (1963). A possible link between active transport of electrolytes and nonelectrolytes. *Fedn Proc. Fedn. Am. Socs exp. Biol.* **22,** 3–7.

FAUST, R. G. and SHEARIN, S. J. (1974). Molecular weight of D-glucose and L-histidine binding protein from intestinal brush border. *Nature, Lond.* **248,** 60–1.

FINEAN, J. B. (1972). The development of ideas on membrane structure. *Sub.-Cell. Biochem.* **1,** 363–73.

FINEAN, J. B., COLEMAN, R. and MICHELL, R. H. (1974). *Membranes and their Cellular Functions.* Blackwell Scientific Publications, Oxford.

FORDTRAN, J. S., RECTOR, F. C. JR. and CARTER, N. W. (1968). The mechanisms of sodium absorption in the human small intestine. *J. clin. Invest.* **47,** 884–900.

FOX, C. F. (1972). The structure of the cell membrane. *Scient. Am.* (Feb. 1972) **226,** 31–8.

HARDCASTLE, P. T. and EGGERTON, J. (1973). The effects of acetylcholine on the electrical activity of intestinal epithelial cells. *Biochim. biophys. Acta* **298,** 95–100.

HEINZ, E. (1967). Transport through biological membranes. *A. Rev. Physiol.* **29,** 21–58.

KIMMICH, G. A. (1970). Active sugar accumulation by isolated intestinal epithelial cells. A new model for sodium-dependent metabolite transport. *Biochemistry, N.Y.* **9**, 3669–77.

LEESE, H. (1973). Ins and outs of the small intestine. *New Scientist* (Nov. 22), 562–64.

LEESE, H. J. and BRONK, J. R. (1972). Glucose accumulation by rat small-intestinal mucosa after depletion of intracellular adenosine triphosphate. *Biochem. J.* **28**, 445–57.

LEVIN, R. J. (1969). *The Living Barrier.* W. Heinemann Medical Books Ltd., London.

MAESTRACCI, D., SCHMITZ, J., PREISER, H. and CRANE, R. K. (1973). Proteins and glycoproteins of the human intestinal brush border membrane. *Biochim. biophys. Acta* **323**, 113–24.

MURER, H. and HOPFER, U. (1974). Demonstration of electrogenic Na^+-dependent D-glucose transport in intestinal brush border membranes. *Proc. natn. Acad. Sci. U.S.A.* **71**, 484–88.

NEAME, K. D. and RICHARDS, T. G. (1972). *Elementary Kinetics of Membrane Carrier Transport.* Blackwell Scientific Publications, Oxford.

OPEN UNIVERSITY TEAM (1974). *Membranes and Transport* (Course S321; units 3 and 4); pp. 1–49. The Open University Press, Milton Keynes.

PARSONS, D. S. and BOYD, C. A. R. (1972). Transport across the intestinal mucosal cell: hierarchies of function. *Int. Rev. Cytol.* **32**, 209–55.

SCHULTZ, S. G. and ZALUSKY, R. (1964). Ion transport in isolated rabbit ileum. II. Interaction between active sodium and active sugar transport. *J. gen. Physiol.* **47**, 1043–59.

SCHULTZ, S. G. and CURRAN, P. F. (1970). Coupled transport of sodium and organic solutes. *Physiol. Rev.* **50**, 637–718.

METHODS

CRANE, R. K. and MANDELSTAM, P. (1960). The active transport of sugars by various preparations of hamster intestine. *Biochem. biophys. Acta* **45**, 460–76.

PARSONS, D. S. (1968). Methods of investigating intestinal absorption. In *Handbook of Physiology.* Ed. C. F. Code. Section 6, Vol. 3, Chapter 64. Pub. Am. Physiol. Soc., Washington, D.C.

SHEFF, M. F. and SMYTH, D. H. (1955). An apparatus for the study of *in vivo* intestinal absorption in the rat. *J. Physiol., Lond.* **128**, 67P.

SMYTH, D. H. (1961). Method for study of intestinal absorption *in vivo. Meth. med. Res.* **9**, 260–72.

WILSON, T. H. and WISEMAN, G. (1954). The use of sacs of everted small intestine for the study of the transference of substances from the mucosal to serosal fluid. *J. Physiol., Lond.* **123**, 116–25.

WISEMAN, G. (1961). Sac of everted intestinal technique for study of intestinal absorption *in vitro. Meth. med. Res.* **9**, 292–97.

CARBOHYDRATES

CRANE, R. K. (1960). Intestinal absorption of sugars. *A. Rev. Physiol.* **40**, 789–825.

CRANE, R. K. (1962). Hypothesis for mechanism of intestinal active transport of sugars. *Fedn Proc. Fedn. Am. Socs exp. Biol.* **21**, 891–95.

CRANE, R. K. (1968). Absorption of sugars. In *Handbook of Physiology.* Ed. C. F. Code. Section 6, Vol. 3, Chapter 69. Pub. Am. Physiol. Soc., Washington, D.C.

GRACEY, M., BURKE, V. and OSHIN, A. (1971). Active intestinal transport of D-fructose. *Biochim. biophys. Acta* **266,** 397–406.

HAMILTON, J. D. and McMICHAEL, H. B. (1968). Role of microvillus in the absorption of disaccharides. *The Lancet* July 28, 1968: 154–7.

MILLER, D. and CRANE, R. K. (1961). The digestive function of the epithelium of the small intestine. *Biochim. biophys. Acta* **52,** 281–98.

ODA, T. and SEKI, S. (1966). Molecular basis of structure and function of the plasma membrane of the microvilli of intestinal epithelial cells. *6th Int. Cong. Electron Microscopy, Kyoto,* 387.

RUBINO, A., ZIMBALATTI, F. and AURICCHIO, S. (1964). Intestinal disaccharidase activities in adult and suckling rats. *Biochim. biophys. Acta* **92,** 305–11.

SCHULTZ, S. G., FUISZ, R. E. and CURRAN, P. F. (1966). Amino acid and sugar transport in rabbit ileum. *J. gen. Physiol.* **49,** 849–66.

SMYTH, D. H. (1971). Sodium-hexose interactions. *Phil. Trans. R. Soc. Ser. B.* **262,** 121–30.

UGOLEV, A. M. (1968). *Physiology and Pathology of Membrane Digestion.* Plenum Press, New York.

UGOLEV, A. M. and DE LACY, P. (1973). Membrane digestion: a concept of enzymatic hydrolysis on cell membranes. *Biochim. biophys. Acta* **300,** 105–28.

AMINO ACIDS AND PROTEINS

CHENG, B., NAVAB, F., LIS, M. T., MILLER, T. N. and MATTHEWS, D. M. (1971). Method of dipeptide uptake by rat small intestine *in vitro. Clin. Sci.* **40,** 247–59.

GROCEY, M. (1973). Immunology and the gut. *Aust. pediat. J.* **9,** 115–20.

GRAY, G. M. and COOPER, H. L. (1971). Protein digestion and absorption. *Gastroenterology,* **61,** 535–44.

JONES, E. D. (1972). Immunoglobulins and the gut. *Gut* **13,** 825–35.

MATTHEWS, D. M. (1972). Rates of peptide uptake by small intestine. In *Peptide transport in Bacteria and Mammalian Gut.* (Eds. Elliott, K. and O'Connor, M.). Ciba Foundation Symposium, pp. 71–92.

MATTHEWS, D. M. (1972). Intestinal absorption of amino acids and peptide. *Proc. Nutr. Soc.* **31,** 171–77.

MILNE, M. D. (1972). Peptides in genetic errors of amino acid transport. In *Peptide transport in Bacteria and Mammalian Gut.* (Eds. Elliott, K. and O'Connor, M.). Ciba Foundation Symposium, pp. 93–106.

MORRIS, I. G. (1968). Gamma globulin absorption in the newborn. In *Handbook of Physiology.* Ed. C. F. Code. Section 6, Vol. 3, Chapter 75. Pub. Am. Physiol. Soc., Washington, D.C.

PETERS, T. J. (1970). The subcellular localization of di- and tri-peptide hydrolase activity in guinea-pig intestine. *Biochem. J.* **120,** 195–203.

ROGERS BRAMBELL, F. W. (1970). *The Transmission of Passive Immunity from Mother to Young.* North-Holland Publishing Company.

SILK, D. B. A. (1974). Peptide absorption in man. *Gut* **15,** 494–501.

SMYTH, D. H. (1961). Intestinal absorption. *Proc. R. Soc. Med.* **54,** 769–79.

WALKER, W. A. (1973). Immunology of the gastrointestinal tract: Part I. *J. Pediat.* **83,** 517–30.

WISEMAN, G. (1968). Absorption of amino acids. In *Handbook of Physiology.* Ed. C. F. Code. Section 6, Vol. 3, Chapter 67. Pub. Am. Physiol. Soc., Washington, D.C.

FATS, CHOLESTEROL AND BILE SALTS

BICKERSTAFFE, R. and ANNISON, E. F. (1960). Triglyceride synthesis by the small intestinal epithelium of the pig, sheep and chicken. *Biochem. J.* **111**, 419–29.

DIETSCHY, J. M. (1968). Mechanisms for the intestinal absorption of bile acids. *J. Lipid. Res.* **9**, 297–309.

DOBBINS III, W. O. (1969). Morphologic aspects of lipid absorption. *Am. J. clin. Nutr.* **22**, 257–65.

FRIEDMAN, H. I. and CARDELL, J. R. (1972). Morphological evidence for the release of chylomicra from intestinal absorptive cells. *Expl Cell Res.* **75**, 57–62.

HOLT, P. R. (1964). Intestinal absorption of bile salts in the rat. *Am. J. Physiol.* **207**, 1–7.

KALIMA, V. K. (1973). Ultrastructure of intestinal lymphatics in regard to absorption. *Scand. J. Gastroent.* **8**, 193–6.

PORTER, K. R. (1969). Independence of fat absorption and pinocytosis. *Fedn Proc. Fedn. Am. Socs exp. Biol.* **28**, 35–40.

STRAUSS, E. W. (1968). Morphological aspects of triglyceride synthesis. In *Handbook of Physiology*. Ed. C. F. Code. Section 6, Vol. 3, Chapter 71. Pub. Am. Physiol. Soc., Washington, D.C.

WATER AND ELECTROLYTES

BARRY, B. A., MATTHEWS, J. and SMYTH, D. H. (1961). Transfer of glucose and fluid by different parts of the small intestine of the rat. *J. Physiol., Lond.* **157**, 279–88.

CAMPEN, D. VAN (1974). Regulation of iron absorption. *Fedn Proc. Fedn. Am. Socs exp. Biol.* **33**, 100–5.

CURRAN, P. F. (1965). Ion transport in intestine and its coupling to other transport processes. *Fedn Proc. Fedn. Am. Socs exp. Biol.* **24**, 993–99.

CURRAN, P. F. (1968). Water absorption from intestine. *Am. J. clin. Nutr.* **21**, 781–84.

CURRAN, P. F. and MacINTOCH, J. R. (1962). A model system for biological water transport. *Nature, Lond.* **193**, 347–48.

DETHERIDGE, J. K., MATTHEWS, J. and SMYTH, D. H. (1966). The effect of inhibitors on intestinal transfer of glucose and fluid. *J. Physiol., Lond.* **183**, 369–77.

DIAMOND, J. (1971). Standard-gradient model of fluid transport in epithelia. *Fedn Proc. Fedn. Am. Socs exp. Biol.* **30**, 6–13.

DIAMOND, J. M. and TORMEY, J. McD. (1966). Role of long extracellular channels in fluid transport across epithelia. *Nature, Lond.* **210**, 817–20.

DIETSCHY, J. M. (1966). Recent developments in solute and water transport across the gall bladder epithelium. *Gastroenterology* **50**, 692–707.

DURBIN, R. P. (1960). Osmotic flow of water across permeable cellulose membranes. *J. gen. Physiol.* **44**, 315–26.

HALJAMÄE, H., JODAL, M. and LUNDGREN, O. (1973). Countercurrent multiplication of sodium in intestinal villi during absorption of sodium chloride. *Acta physiol. scand.* **89**, 580–94.

HEINZ, E. ed. (1972). *Na⁺-linked Transport of Organic Solutes*. Springer-Verlag, Berlin.

KEYNES, R. D. (1969). From frog skin to sheep rumen: a survey of transport of salts and water across multicellular structures. *Q. Rev. Biophys.* **2**, 177–281.

KIMMICH, G. A. (1973). Coupling between Na⁺ and sugar transport in small intestine. *Biochim. biophys. Acta* **300**, 31–78.

LEE, J. F. (1961). Flows and pressures in lymphatic and blood vessels of intestine in water absorption. *Am. J. Physiol.* **200,** 979–83.

MADGE, D. S. (1972). Fluid and glucose movement across the small intestine in laboratory animals. *Comp. Biochem. Physiol.* **43A,** 565–75.

SCHEDL, H. P. (1966). Present concepts of water and electrolyte transport. *Gastroenterology,* **50,** 134–37.

SCHULTZ, S. G. and FIZZELL, R. A. (1972). An overview of intestinal absorption and secretory processes. *Gastroenterology,* **63,** 161–170.

SCHULTZ, S. G., FIZZELL, R. A. and NELLANS, H. N. (1974). Ion transport by mammalian small intestine. *A. Rev. Physiol.* **36,** 51–91.

STEVEN, D. H. and MARSHALL, A. B. (1970). Organisation of the rumen epithelium. *In Physiology of Digestion and Metabolism in the Ruminant.* Ed. A. T. Phillipson. Oriel Press, Newcastle upon Tyne, pp. 80–100.

TURNBERG, L. A. (1973). Absorption and secretion of salt and water by the small intestine. *Digestion* **9,** 357–81.

VITAMINS

HERBERT, V. (1968). Absorption of vitamin B_{12} and folic acid. *Gastroenterology* **54,** 110–15.

MACKENZIE, I. L. and DONALDSON, R. M. JR.(1969). Vitamin B_{12} absorption and the intestinal cell surface. *Fedn Proc. Fedn. Am. Socs exp. Biol.* **28,** 41–5.

STRAUSS, E. W. and WILSON, T. H. (1960). Factors controlling B_{12} uptake by intestinal sacs *in vitro. Am. J. Physiol.* **198,** 103–7.

MALABSORPTION

BOGOCH, E. ed. (1973). *Gastroenterology.* Chapters 12, 13, 18 and 19. McGraw-Hill Book Co., New York.

CREAMER, B. (1974). *The Small Intestine.* Chapters 4–17. W. Heinemann Medical Books Ltd., London.

DAWSON, A. M. ed. (1971). Intestinal Absorption and its Derangements. *J. clin. Path.* **24,** Suppl. (Roy. Coll. Path.) **5.**

GILLESPIE, I. E. and THOMSON, T. J. (1972). *Gastroenterology. An Integrated Approach.* Chapters 6 and 8. Churchill Livingstone, London.

GRADY, G. F. and KEUSCH, T. G. (1971). Pathogenesis of bacterial diarrheas. *New Eng. J. Med.* **285,** 831–41, 891–900.

LOW-BEER, T. S. and READ, A. E. (1972). Diarrhoea: mechanism and treatment. *Gut* **12,** 1021–36.

MADGE, D. S. (1973). Small intestine carbohydrase activities in experimental *Salmonella* enterocolitis in mice. *Life Sci.* **12,** 535–42.

MADGE, D. S. (1972). Intestinal absorption in experimental *Salmonella* enterocolitis in mice. *Comp. Biochem. Physiol.* **42A,** 445–63.

MORSON, B. C. and DAWSON, I. M. P. (1972). *Gastrointestinal Pathology.* Chapters 20 and 21. Blackwell Scientific Publications, Oxford.

PHILLIPS, S. F. (1972). Diarrhea: a current view of the pathophysiology. *Gastroenterology* **63,** 495–518.

SOULSBY, E. J. L. (1965). *Textbook of Veterinary Clinical Pathology. Volume I. Helminths.* Blackwell Scientific Publications, Oxford.

SYMONS, L. E. A. (1969). Pathology of gastrointestinal helminths. *Int. Rev. Trop. Med.* **39,** 49–100.

Index

Index